DEFENSE, SECURITY AND STRATEGIES

ELECTRONIC WARFARE

SELECT ASSESSMENTS OF MILITARY CAPABILITIES AND STRATEGIES

DEFENSE, SECURITY AND STRATEGIES

Additional books in this series can be found on Nova's website under the Series tab.

Additional E-books in this series can be found on Nova's website under the E-book tab.

AMERICAN POLITICAL, ECONOMIC, AND SECURITY ISSUES

Additional books in this series can be found on Nova's website under the Series tab.

Additional E-books in this series can be found on Nova's website under the E-book tab.

DEFENSE, SECURITY AND STRATEGIES

ELECTRONIC WARFARE

SELECT ASSESSMENTS OF MILITARY CAPABILITIES AND STRATEGIES

MASON BOYD
AND
PIERCE T. CHAPMAN
EDITORS

nova publishers
New York

Library of Congress Cataloging-in-Publication Data

ISBN: 978-1-62257-792-7

Published by Nova Science Publishers, Inc. † New York

CONTENTS

PREFACE

According to the DOD, the U.S. military's access to and use of the electromagnetic spectrum is facing rapidly evolving challenges and increased vulnerabilities due to the increasing quality and availability of electronic warfare capabilities to both state and non-state actors. Also, DOD has reported that electronic warfare capabilities, which play a critical and potentially growing role as an enabler for military operations, are currently stressed and will remain so in the future. Moreover, near-peer competitors, primarily Russia and China, have fully recognized the critical nature of electromagnetic spectrum control in military operations and cyperspace operations, which includes computer network operations. This book examines select assessments of military capabilities and strategies with regard to electronic warfare, with a focus on management and oversight, as well as airborne electronic attacks.

Chapter 1 - DOD has committed billions of dollars to developing, maintaining, and employing warfighting capabilities that rely on access to the electromagnetic spectrum. According to DOD, electronic warfare capabilities play a critical and potentially growing role in ensuring the U.S. military's access to and use of the electromagnetic spectrum. GAO was asked to assess the extent to which DOD (1) developed a strategy to manage electronic warfare and (2) planned, organized, and implemented an effective governance structure to oversee its electronic warfare policy and programs and their relationship to cyberspace operations. GAO analyzed policies, plans, and studies related to electronic warfare and cyberspace operations and interviewed cognizant DOD officials.

Chapter 2 - Airborne electronic attack involves the use of aircraft to neutralize, destroy, or suppress enemy air defense and communications systems. Proliferation of sophisticated air defenses and advanced commercial electronic devices has contributed to the accelerated appearance of new weapons designed to counter U.S. airborne electronic attack capabilities. GAO was asked to assess (1) the Department of Defense's (DOD) strategy for acquiring airborne electronic attack capabilities, (2) progress made in developing and fielding systems to meet airborne electronic attack mission requirements, and (3) additional actions taken to address capability gaps. To do this, GAO analyzed documents related to mission requirements, acquisition and budget needs, development plans, and performance, and interviewed DOD officials.

In: Electronic Warfare
Editors: M. Boyd and P.T. Chapman
ISBN: 978-1-62257-792-7

Chapter 1

ELECTRONIC WARFARE: DOD ACTIONS NEEDED TO STRENGTHEN MANAGEMENT AND OVERSIGHT*

United States Government Accountability Office

WHY GAO DID THIS STUDY

DOD has committed billions of dollars to developing, maintaining, and employing warfighting capabilities that rely on access to the electromagnetic spectrum. According to DOD, electronic warfare capabilities play a critical and potentially growing role in ensuring the U.S. military's access to and use of the electromagnetic spectrum.

GAO was asked to assess the extent to which DOD (1) developed a strategy to manage electronic warfare and (2) planned, organized, and implemented an effective governance structure to oversee its electronic warfare policy and programs and their relationship to cyberspace operations. GAO analyzed policies, plans, and studies related to electronic warfare and cyberspace operations and interviewed cognizant DOD officials.

---

* This is an edited, reformatted and augmented version of the Highlights of GAO-12-479, a report to the Committee on Armed Services, House of Representatives, dated July 2012.

WHAT GAO RECOMMENDS

GAO recommends that DOD should (1) include in its future electronic warfare strategy reports to Congress certain key characteristics, including performance measures, key investments and resources, and organizational roles and responsibilities; (2) define objectives and issue an implementation plan for the Joint Electromagnetic Spectrum Control Center; and (3) update key departmental guidance to clearly define oversight roles, responsibilities, and coordination for electronic warfare management, and the relationship between electronic warfare and cyberspace operations. DOD generally concurred with these recommendations, except that the strategy should include performance measures. GAO continues to believe this recommendation has merit.

WHAT GAO FOUND

The Department of Defense (DOD) developed an electronic warfare strategy, but it only partially addressed key characteristics that GAO identified in prior work as desirable for a national or defense strategy. The National Defense Authorization Act for Fiscal Year 2010 requires DOD to submit to the congressional defense committees an annual report on DOD's electronic warfare strategy for each of fiscal years 2011 through 2015. DOD issued its fiscal year 2011 and 2012 strategy reports to Congress in October 2010 and November 2011, respectively. GAO found that DOD's reports addressed two key characteristics: (1) purpose, scope, and methodology and (2) problem definition and risk assessment.

However, DOD only partially addressed four other key characteristics of a strategy, including (1) resources, investments, and risk management and (2) organizational roles, responsibilities, and coordination.

For example, the reports identified mechanisms that could foster coordination across the department and identified some investment areas, but did not fully identify implementing parties, delineate roles and responsibilities for managing electronic warfare across the department, or link resources and investments to key activities.

Such characteristics can help shape policies, programs, priorities, resource allocation, and standards in a manner that is conducive to achieving intended results and can help ensure that the department is effectively managing electronic warfare.

DOD has taken steps to address a critical electronic warfare management gap, but it has not established a departmentwide governance framework for electronic warfare. GAO previously reported that effective and efficient organizations establish objectives and outline major implementation tasks. In response to a leadership gap for electronic warfare, DOD is establishing the Joint Electromagnetic Spectrum Control Center under U.S. Strategic Command as the focal point for joint electronic warfare. However, because DOD has yet to define specific objectives for the center, outline major implementation tasks, and define metrics and timelines to measure progress, it is unclear whether or when the center will provide effective departmentwide leadership and advocacy for joint electronic warfare.

In addition, key DOD directives providing some guidance for departmentwide oversight of electronic warfare have not been updated to reflect recent changes. For example, DOD's primary directive concerning electronic warfare oversight was last updated in 1994 and identifies the Under Secretary of Defense for Acquisition, Technology, and Logistics as the focal point for electronic warfare. The directive does not define the center's responsibilities in relation to the office, including those related to the development of the electronic warfare strategy and prioritizing investments. In addition, DOD's directive for information operations, which is being updated, allocates electronic warfare responsibilities based on the department's previous definition of information operations, which had included electronic warfare as a core capability.

DOD's oversight of electronic warfare capabilities may be further complicated by its evolving relationship with computer network operations, which is also an information operations-related capability. Without clearly defined roles and responsibilities and updated guidance regarding oversight responsibilities, DOD does not have reasonable assurance that its management structures will provide effective departmentwide leadership for electronic warfare activities and capabilities development and ensure effective and efficient use of its resources.

ABBREVIATIONS

DOD	Department of Defense
JEMSCC	Joint Electromagnetic Spectrum Control Center

July 9, 2012

The Honorable Howard P. "Buck" McKeon
Chairman

The Honorable Adam Smith
Ranking Member
Committee on Armed Services
House of Representatives

The Department of Defense (DOD) is increasingly dependent on access to the electromagnetic spectrum—the full range of all possible frequencies of electromagnetic radiation, including frequency ranges such as radio, microwave, infrared, visible, ultraviolet, X-rays, and gamma rays—for a variety of military uses, such as communicating, navigating, information gathering and sensing, and targeting.

DOD has committed billions of dollars developing, maintaining, and employing warfighting capabilities that rely on access to the electromagnetic spectrum—including precision-guided munitions and command, control, and communications systems. DOD ensures control of the electromagnetic spectrum through the coordinated implementation of joint electromagnetic spectrum operations, which includes electronic warfare and spectrum management activities, with other lethal and nonlethal operations that enable freedom of action in the electromagnetic operational environment.

Electronic warfare, which is the use of electromagnetic energy and directed energy to control the electromagnetic spectrum or to attack the enemy, is essential for protection of friendly operations and denying adversary operations within the electromagnetic spectrum throughout the operational environment. As we previously reported, DOD's investments are projected to total more than $17.6 billion from fiscal years 2007 through 2016 for the development and procurement of new and updated fixed-wing airborne electronic attack systems alone, which are one element of electronic warfare.[1]

According to DOD, the U.S. military's access to and use of the electromagnetic spectrum is facing rapidly evolving challenges and increased vulnerabilities due to the increasing quality and availability of electronic warfare capabilities to both state and non-state actors. Also, DOD has reported that electronic warfare capabilities, which play a critical and potentially growing role as an enabler for military operations, are currently stressed and will remain so in the future.

Moreover, according to DOD, near-peer competitors, primarily Russia and China, have fully recognized the critical nature of electromagnetic spectrum control in military operations.[2] There also has been recognition among near-peer competitors of the relationship between electronic warfare and cyberspace operations, which includes computer network operations.[3] For example, as noted in the U.S.-China Economic and Security Review Commission's 2009 report to Congress, China's Integrated Network Electronic Warfare concept incorporates elements of cyberspace operations in tandem with elements of traditional electronic warfare, and advocates for the employment of traditional electronic warfare operations—such as the jamming of radars and communications systems—in coordination with cyberspace attack operations.

DOD has identified persistent electronic warfare capability gaps, and these shortfalls have been consistently highlighted by the combatant commands as some of their highest warfighting priorities. According to a Center for Strategic and International Studies report, the U.S. Strategic Command identified 34 capability gaps affecting electronic warfare, including a lack of leadership across the department.[4] This lack of leadership was identified as the most critical gap. In our recent report on DOD's airborne electronic attack capabilities, we found that DOD is developing multiple systems which provide similar capabilities, and that the lack of leadership may undermine DOD's ability to consolidate these systems.[5]

Specifically, we found that all four military services within the Department of Defense are separately acquiring new airborne electronic attack systems, but that opportunities may exist to consolidate some current service-specific acquisition efforts. With the prospect of slowly growing or flat defense budgets for years to come, the department must get better returns on its weapon system investments and find ways to deliver more capability to the warfighter for less than it has in the past.

Therefore, we recommended that the Secretary of Defense conduct program reviews for certain new, key systems; determine the extent to which the most pressing capability gaps can be met and take steps to fill them; align service investments in science and technology with the departmentwide electronic warfare priority; and review the capabilities provided by certain existing and planned systems to ensure investments do not overlap. DOD generally concurred with our recommendations.

You requested that we examine several issues related to DOD's electronic warfare capabilities. In March 2012, we issued a report on DOD's current and planned airborne electronic attack capabilities and investment strategies.[6] In this current review, we examined DOD's approach to governing electronic warfare and the relationship between electronic warfare and cyberspace operations. Specifically, we examined the extent to which DOD has (1) developed a strategy to manage electronic warfare and (2) planned, organized, and implemented an effective governance structure to oversee its electronic warfare policy and programs, and their relationship to cyberspace operations.

To assess the extent to which DOD has developed a strategy to manage electronic warfare, we compared information found in DOD's two electronic warfare strategy reports to Congress with key characteristics of strategies identified by GAO in prior work, and interviewed relevant officials. To assess the extent to which DOD has planned, organized, and implemented an effective governance structure to oversee its electronic warfare policy and programs and their relationship to cyberspace operations, we reviewed DOD directives and policies, and the roles and responsibilities of the Under Secretary of Defense for Policy; Under Secretary of Defense for Acquisition, Technology, and Logistics; and U.S. Strategic Command. Additionally, we reviewed and analyzed information found in policy documents along with information from relevant meetings with DOD officials against DOD's directives regarding electronic warfare.

We also interviewed cognizant officials and reviewed DOD policies, doctrine, reports, plans, and concepts of operation, and outside studies that discuss the relationship between electronic warfare and cyberspace operations. See Appendix I for details on our scope and methodology.

We conducted this performance audit from July 2011 to July 2012 in accordance with generally accepted government auditing standards. Those standards require that we plan and perform the audit to obtain sufficient, appropriate evidence to provide a reasonable basis for our findings and conclusions based on our audit objectives. We believe that the evidence obtained provides a reasonable basis for our findings and conclusions based on our audit objectives.

BACKGROUND

Control and Use of the Electromagnetic Spectrum

In modern warfare, military forces are heavily dependent upon access to the electromagnetic spectrum for successful operations. Communications with friendly forces and detection, identification, and targeting of enemy forces, among other tasks, are all reliant upon the ability to operate unhindered in the spectrum. For this reason, control of the electromagnetic spectrum is considered essential to carrying out military operations.[7] Figure 1 illustrates the electromagnetic spectrum and some examples of military uses at various frequencies. For example, infrared or thermal imaging technology senses heat emitted by a person or an object and creates an image. Sensor systems utilize this technology to provide the advantage of seeing not only at night but also through smoke, fog, and other obscured battlefield conditions.

Electronic Warfare

DOD defines electronic warfare as any military action involving the use of electromagnetic and directed energy to control the electromagnetic spectrum or to attack the enemy. The purpose of electronic warfare is to secure and maintain freedom of action in the electromagnetic spectrum for friendly forces and to deny the same for the adversary. Traditionally, electronic warfare has been composed of three primary activities:

- Electronic attack: use of electromagnetic, directed energy, or antiradiation weapons to attack personnel, facilities, or equipment with the intent of degrading, neutralizing, or destroying enemy combat capability. Electronic attack can be used offensively, such as jamming enemy communications or jamming enemy radar to suppress its air defenses, and defensively, such as deploying flares.
- Electronic protection: actions to protect personnel, facilities, and equipment from any effects of friendly, neutral, or enemy use of the electromagnetic spectrum, as well as naturally occurring phenomena that degrade, neutralize, or destroy friendly combat capability.
- Electronic warfare support: actions directed by an operational commander to search for, intercept, identify, and locate sources of radiated electromagnetic energy for the purposes of immediate threat recognition, targeting, and planning; and conduct of future operations.

Information Operations

Electronic warfare is employed to create decisive stand-alone effects or to support military operations, such as information operations and cyberspace operations. According to DOD, information operations are the integrated employment, during military operations, of information-related capabilities in concert with other lines of operation to influence, disrupt, corrupt, or usurp the decision-making of adversaries and potential adversaries while protecting our own. Information-related capabilities can include, among others, electronic warfare, computer network operations, military deception, operations security, and military information support operations (formerly psychological operations). Electronic warfare contributes to the success of information operations by using offensive and defensive tactics and techniques in a variety of combinations to shape, disrupt, and exploit adversarial use of the electromagnetic spectrum while protecting U.S. and allied freedom of action.

Cyberspace Operations

Since cyberspace requires both wired and wireless links to transport information, both offensive and defensive cyberspace operations may require use of the electromagnetic spectrum.

According to DOD, cyberspace operations are the employment of cyberspace capabilities where the primary purpose is to achieve military objectives or effects through cyberspace, which include computer network operations, among others.

Computer network operations include computer network attack, computer network defense, and related computer network exploitation-enabling operations. Electronic warfare and cyberspace operations are complementary and have potentially synergistic effects. For example, an electronic warfare platform may be used to enable or deter access to a computer network.

U.S. Strategic Command Joint Electronic Warfare Activities

U.S. Strategic Command (Strategic Command) has been designated since 2008 as the advocate for joint electronic warfare. Strategic Command officials stated that, in the past, the primary office for electronic warfare expertise—the Joint Electronic Warfare Center—had several different names and was aligned under several different organizations, such as the Joint Forces Command and the U.S. Space Command.

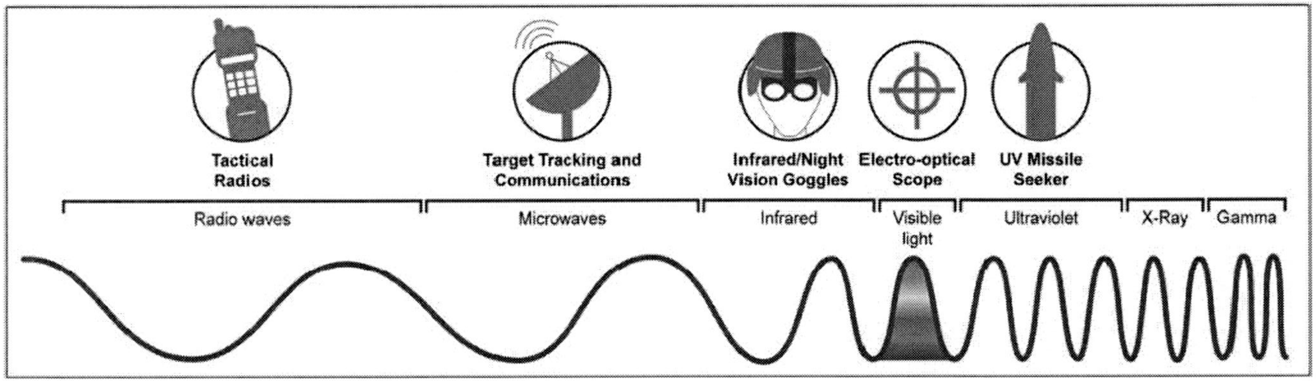

Source: GAO analysis based on DOD information.

Figure 1. Electromagnetic Spectrum and Uses.

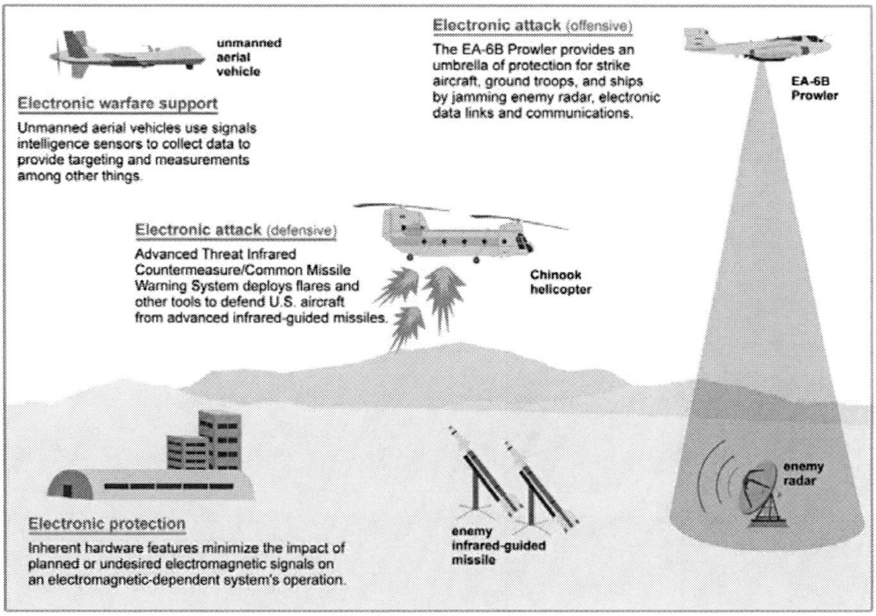

Source: GAO analysis based on DOD information.

Figure 2. Examples of Electronic Warfare Capabilities.

According to Strategic Command officials, in addition to the Joint Electronic Warfare Center, the command employs electronic warfare experts in its non-kinetic operations staff and in the Joint Electromagnetic Preparedness for Advanced Combat organization. According to Strategic Command officials, the Joint Electronic Warfare Center is the largest of the three organizations and employs approximately 60 military and civilian electronic warfare personnel and between 15 and 20 contractors. Strategic Command officials stated that the Joint Electronic Warfare Center was created as a DOD center of excellence for electronic warfare and has electronic warfare subject matter experts. The center provides planning and technical support not only to Strategic Command but to other combatant commands and organizations, such as U.S. Central Command, U.S. European Command, U.S. Pacific Command, and the Department of Homeland Security. The Joint Electronic Warfare Center also provides assistance with requirements generation to the military services.

Desirable Characteristics and Summary Analyses	GAO assessment
Purpose, scope, and methodology	
Addressed	●
Problem definition and risk assessment	
Addressed	●
Goals, subordinate objectives, activities, and performance measures	
Partially addressed—The report identifies the goal, objectives, and several activities associated with the strategy, but does not identify outcome-related performance measures that could help ensure accountability.	◒
Resources, investments, and risk management	
Partially addressed—The report refers to investment areas and provides data on acquisition programs and research and development projects, but does not target resources and investments at key activities associated with implementing the strategy, or balance risk against cost.	◒
Organizational roles, responsibilities, and coordination	
Partially addressed—The report provides an overview of some electronic warfare activities and identifies mechanisms that could be used to foster coordination across the department, but it does not fully identify which entities will be implementing the strategy or discuss implementing entities' roles and responsibilities.	◒
Integration and implementation	
Partially addressed—The report describes how the strategy supports the objectives and activities of the National Defense Strategy, but does not address implementation plans for the strategy.	◒

● Addressed
◒ Partially addressed
○ Does not address

Source: GAO analysis of DOD's fiscal year 2011 *Electronic Warfare Strategy of the Department of Defense* report to Congress.

Notes: Summary analysis information is provided only in cases where we have determined the strategy partially addressed a characteristic. The strategy "addressed" a characteristic when the strategy explicitly cited all elements of a characteristic, even if it lacked specificity and details and thus could be improved upon. The strategy "partially addressed" a characteristic when the strategy explicitly cited some, but not all, elements of a characteristic. Within our designation of "partially addressed," there may be wide variation between a characteristic for which most of the elements were addressed and a characteristic for which few of the elements of the characteristic were addressed. The strategy "did not address" a characteristic when the strategy did not explicitly cite or discuss any elements of a characteristic, and/or any implicit references were either too vague or general.

Figure 3. Extent to Which DOD's Fiscal Year 2011 Electronic Warfare Strategy Report Addressed Key Desirable Strategy Characteristics Identified by GAO.

DOD DEVELOPED AN ELECTRONIC WARFARE STRATEGY, BUT ONLY PARTIALLY ADDRESSED KEY DESIRABLE STRATEGY CHARACTERISTICS

DOD developed an electronic warfare strategy, but only partially addressed key strategy characteristics identified as desirable in prior work by GAO. The *National Defense Authorization Act for Fiscal Year 2010* requires the Secretary of Defense to submit to the congressional defense committees an annual report on DOD's electronic warfare strategy for each of fiscal years 2011 through 2015.[8] Each annual report is to be submitted at the same time the President submits the budget to Congress and is to contain, among other things, a description and overview of DOD's electronic warfare strategy and the organizational structure assigned to oversee the development of the department's electronic warfare strategy, requirements, capabilities, programs, and projects.[9] In response to this legislative requirement, the Office of the Under Secretary of Defense for Acquisition, Technology, and Logistics issued DOD's 2011 and 2012 fiscal year strategy reports to Congress in October 2010 and November 2011, respectively.[10]

We previously reported that it is desirable for strategies to delineate six key characteristics, including organizational roles and responsibilities for implementing parties as well as performance measures to gauge results.[11] The key characteristics of an effective strategy can aid responsible parties in further developing and implementing the strategy, enhance the strategy's usefulness in resource and policy decisions, and better ensure accountability. The six characteristics are: (1) purpose, scope, and methodology; (2) problem definition and risk assessment; (3) goals, subordinate objectives, activities, and performance measures; (4) resources, investments, and risk management; (5) organizational roles, responsibilities, and coordination; and (6) integration and implementation.

As illustrated in Figure 3, we found that DOD's reports addressed two key characteristics, but only partially addressed four other key characteristics of a strategy. For example, the strategy reports to Congress included elements of characteristics, such as a goal and objectives, but did not fully identify implementing parties, delineate roles and responsibilities for managing electronic warfare across the department, or identify outcome-related performance measures that could guide the implementation of electronic warfare efforts and help ensure accountability. Similarly, the reports provided acquisition program and research and development project data, but did not target resources and investments at some key activities associated with implementing the strategy. When investments are not tied to strategic goals and priorities, resources may not be used effectively and efficiently. Our past work has shown that such characteristics can help shape policies, programs, priorities, resource allocations, and standards in a manner that is conducive to achieving intended results.[12]

DOD's fiscal year 2011 report is described here because the fiscal year 2012 report, issued in November 2011, is classified. However, unclassified portions of this document note that the fiscal year 2011 report remains valid as the base DOD strategy and that the fiscal year 2012 report updates its predecessor primarily to identify ongoing efforts to improve DOD's electronic warfare capabilities and to provide greater specificity to current threats. The fiscal year 2011 *Electronic Warfare Strategy of the Department of Defense* report (electronic warfare strategy report)—the base electronic warfare strategy—addressed two and partially addressed four of six desirable characteristics of a strategy identified by GAO. There may be considerable variation in the extent to which the strategy addressed specific elements of those characteristics that were determined by GAO to be partially addressed. Our analysis of the fiscal year 2011 report's characteristics is as follows.

- **Purpose, scope and methodology:** ***Addressed.*** The fiscal year 2011 electronic warfare strategy report identifies the purpose of the strategy, citing as its impetus section 1053 of the *National Defense Authorization Act for Fiscal Year 2010*, and articulates a maturing, twofold strategy focused on integrating electronic warfare capabilities into all phases and at all levels of military operations, as well as developing, maintaining, and protecting the maneuver space within the electromagnetic spectrum necessary to enable military capabilities. The report's scope also encompasses data on acquisition programs and research and development projects. Additionally, the report includes some methodological information by citing a principle that guided its development. Specifically the report states that a key aspect of the strategy is the concept of the electromagnetic spectrum as maneuver space.
- **Problem definition and risk assessment:** ***Addressed.*** The fiscal year 2011 electronic warfare strategy report defines the problem the strategy intends to address, citing the challenges posed to U.S. forces by potential adversaries' increasingly sophisticated technologies, the military's increased dependence on the electromagnetic spectrum, and

the urgent need to retain and expand remaining U.S. advantages. The report also assesses risk by identifying threats to, and vulnerabilities of critical operations, such as Airborne Electronic Attack and self-protection countermeasures.

- **Goals, subordinate objectives, activities, and performance measures: *Partially Addressed.*** The fiscal year 2011 electronic warfare strategy report communicates an overarching goal of enabling electromagnetic spectrum maneuverability and cites specific objectives, such as selectively denying an adversary's use of the spectrum and preserving U.S. and allied forces' ability to maneuver within the spectrum. The report also identifies key activities associated with the strategy, including developing (1) coherent electronic warfare organizational structures and leadership, (2) an enduring and sustainable approach to continuing education, and (3) capabilities to implement into electronic warfare systems. The report does not identify performance measures that could be used to gauge results and help ensure accountability.
- **Resources, investments, and risk management: *Partially Addressed.*** The fiscal year 2011 electronic warfare strategy report broadly targets resources and investments by emphasizing the importance of continued investment in electronic attack, electronic protection, and electronic support capabilities. The report also notes some of the associated risks in these areas, calling for new methods of ensuring U.S. control over the electromagnetic spectrum in light of the adversary's advances in weapons and the decreasing effectiveness of traditional lines of defense, such as airborne electronic attack and self-protection countermeasures. The report identifies some of the costs associated with the strategy by providing acquisition program and research and development project and cost data, and notes that part of the strategy is to identify and track investments in electronic warfare systems, which often are obscured within the development of the larger weapons platforms they typically support. However, the strategy does not target investments by balancing risk against costs, or discuss other costs associated with implementing the strategy by, for example, targeting resources and investments at key activities, such as developing electronic warfare organizational structures and leadership and developing an enduring and sustainable approach to continuing education.

- **Organizational roles, responsibilities, and coordination: *Partially Addressed.*** The fiscal year 2011 electronic warfare strategy report provides an overview of past and ongoing electronic warfare activities within the military services and DOD, and identifies several mechanisms that have or could be used to foster coordination across the department. For example, it outlines the Army's efforts to create a new career field for electronic warfare officers and the Office of the Under Secretary of Defense for Acquisition, Technology and Logistics' electronic warfare integrated planning team.[13] However, the report does not fully identify the departmental entities responsible for implementing the strategy, discuss the roles and responsibilities of implementing parties, or specify implementing entities' relationships in terms of leading, supporting, and partnering.[14]
- **Integration and implementation: *Partially Addressed.*** The fiscal year 2011 electronic warfare strategy report describes the department's approach to ensuring maneuverability within the electromagnetic spectrum, thus supporting *National Defense Strategy* objectives that rely on use and control of the spectrum. The strategy's overarching aim of ensuring electromagnetic spectrum maneuverability also is consistent with concepts contained in the department's electromagnetic spectrum strategy documents—which collectively emphasize the importance of assured spectrum access.[15] The strategy does not, however, discuss the department's plans for implementing the strategy.

DOD's electronic warfare strategy reports were issued in response to the *National Defense Authorization Act for Fiscal Year 2010* and were not specifically required to address all the characteristics we consider to be desirable for an effective strategy. Additionally, DOD's fiscal year 2011 report states that the strategy is still maturing and that subsequent reports to Congress will refine the department's vision. Nonetheless, we consider it useful for DOD's electronic warfare strategy to address each of the characteristics we have identified in order to provide guidance to the entities responsible for implementing DOD's strategy and to enhance the strategy's utility in resource and policy decisions—particularly in light of the diffuse nature of DOD's electronic warfare programs and activities, as well as the range of emerging technical, conceptual, and organizational challenges and changes in this area. Further, in the absence of clearly defined roles and responsibilities, and other elements of key characteristics, such as measures of performance in meeting goals and objectives, entities responsible for implementing DOD's strategy may lack the guidance necessary to establish priorities and milestones, thereby impeding their ability to achieve intended results within a reasonable time frame. As a result, DOD lacks assurance that its electronic warfare programs and activities are aligned with strategic priorities and are managed effectively. For example, without an effective strategy, DOD is limited in its ability to reduce the potential for unnecessary overlap in the airborne electronic attack acquisition activities on which we have previously reported.

DOD Has Not Established an Effective Departmentwide Governance Framework for Managing and Overseeing Electronic Warfare

DOD has taken some steps to address a critical leadership gap identified in 2009, but it has not established a departmentwide governance framework for planning, directing, and controlling electronic warfare activities. DOD is establishing a Joint Electromagnetic Spectrum Control Center (JEMSCC) under Strategic Command in response to the leadership gap for electronic warfare. However, DOD has not documented the objectives or implementation tasks and timeline for the JEMSCC. In addition, DOD has not updated key guidance to reflect recent policy changes regarding electronic warfare management and oversight roles and responsibilities. For example, it is unclear what the JEMSCC's role is in relation to other DOD organizations involved in the management of electronic warfare, such as the Office of the Under Secretary of Defense for Acquisition, Technology, and Logistics. Moreover, we found that DOD may face challenges in its oversight of electronic warfare as a result of the evolving relationship between electronic warfare and cyberspace operations.

DOD Actions Have Not Fully Addressed a Critical Leadership Gap

DOD has taken some steps to address a critical leadership gap by establishing the JEMSCC under Strategic Command. However, because DOD has yet to define specific objectives for the center, outline major implementation tasks, and define metrics and timelines to measure progress, it is unclear to what extent the center will address the identified existing leadership deficiencies. The Center for Strategic and International Studies reported insufficient leadership as the most critical among 34 capability gaps affecting electronic warfare. As a result of the absence of leadership, the department was significantly impeded from both identifying departmentwide needs and solutions and eliminating potentially unnecessary overlap among the military services' electronic warfare acquisitions. Specifically, the department lacked a joint leader and advocate with the authority to integrate and influence electronic warfare capabilities development, to coordinate internal activities, and to represent those activities and interests to outside organizations. Mitigating the leadership gap was identified not only as the highest priority, but also a prerequisite to addressing the other 33 gaps.

The Center for Strategic and International Studies report was one of two parallel studies commissioned by the Joint Requirements Oversight Council[16] to assess potential organizational and management solutions to the leadership gap.[17] These studies considered a number of options, including an organization under the Deputy Secretary of Defense, an activity controlled by the Chairman of the Joint Chiefs of Staff, and an organization at Strategic Command. As a result of these studies, in January 2011, DOD initiated efforts to establish the JEMSCC under Strategic Command as the focal point of joint electronic warfare advocacy. This solution was chosen, in part, in recognition of Strategic Command's resident electronic warfare expertise as well as its already assigned role as an electronic warfare advocate.[18]

In January 2011, the Joint Requirements Oversight Council directed Strategic Command to develop an implementation plan for the electronic warfare center to be submitted for council approval no later than May 2011. The plan was to delineate (1) the center's mission, roles, and responsibilities; (2) command and control, reporting, and support relationships with combatant commands, military services, and U.S. Government departments and agencies; and (3) minimum requirements to achieve initial operational capability and full operational capability. The Joint Requirements Oversight Council subsequently approved an extension of the center's implementation plan submission to August 2011. Subsequently, in December 2011, the oversight council issued a memorandum that closed the requirement to submit an implementation plan to the council and stated that Strategic Command had conducted an internal reorganization and developed a center to perform the functions identified in the internal DOD study.

In December 2011, Strategic Command issued an operations order that defined the JEMSCC as the primary focal point for electronic warfare, supporting DOD advocacy for joint electronic warfare capability requirements, resources, strategy, doctrine, planning, training, and operational support. This order provided 22 activities that the center is to perform. Federal internal control standards require that organizations establish objectives and clearly define key areas of authority and responsibility.[19] In addition, best practices for strategic planning have shown that effective and efficient operations require detailed plans outlining major implementation tasks and defined metrics and timelines to measure progress.[20] Moreover, the independent study prepared for DOD similarly emphasized the importance of clearly defining the center's authorities and responsibilities, noting that the center's success would hinge, in part, on specifying how it is expected to relate to the department as a whole as well as its expected organizational outcomes. However, as of March 2012, Strategic Command had not issued an implementation plan or other documentation that defines the center's objectives and outlines major implementation tasks, metrics, and timelines to measure progress. Strategic Command officials told us in February 2012 that an implementation plan had been drafted, but that there were no timelines for the completion of the implementation plan or a projection for when the center would reach its full operational capability. As a result, it remains unclear whether or when the JEMSCC will provide effective departmentwide leadership and advocacy for electronic warfare, and influence resource decisions related to capability development.

According to officials from Strategic Command, the JEMSCC will consist of staff from Strategic Command's Joint Electronic Warfare Center at Lackland Air Force Base, Texas, and the Joint Electromagnetic Preparedness for Advanced Combat organization, at Nellis Air Force Base, Nevada.[21] These officials stated that while each of JEMSCC's component groups' missions will likely evolve as the center matures, the JEMSCC components would continue prior support activities, such as the Joint Electronic Warfare Center's support to other combatant commands through its Electronic Warfare Planning and Coordination Cell—a rapid deployment team that provides electronic warfare expertise and support to build electronic warfare capacity. Figure 4 depicts the JEMSCC's organizational construct.

DOD has yet to define objectives and issue an implementation plan for the JEMSCC; however, officials from Strategic Command stated that they anticipated continuity between the command's previous role as an electronic warfare advocate and its new leadership role, noting that advocacy was, and remains, necessary because electronic warfare capabilities are sometimes undervalued in comparison to other, kinetic capabilities.[22] For example, the

JEMSCC will likely build off Strategic Command's previously assigned advocacy role, in part, by continuing to advocate for electronic warfare via the Joint Capabilities Integration and Development System process—DOD's process for identifying and developing capabilities needed by combatant commanders—and by providing electronic warfare expertise.[23] Specifically, Strategic Command officials stated that the JEMSCC, through Strategic Command, would likely provide input to the development of joint electronic warfare requirements during the joint capabilities development process. However, combatant commands, such as Strategic Command, provide one of many inputs to this process. Further, as we have previously reported, council decisions, while influential, are advisory to acquisition and budget processes driven by military service investment priorities.[24] As a result, the JEMSCC's ability to affect resource decisions via this process is likely to be limited.

Officials we spoke with across DOD, including those from the military services and Strategic Command, recognized this challenge. Specifically, Strategic Command officials told us that for JEMSCC to influence service-level resource decisions and advocate effectively for joint electronic warfare capabilities, the JEMSCC would need to not only participate in the joint capabilities development process, but would also need authorities beyond those provided by the *Unified Command Plan*, such as the authority to negotiate with the military services regarding resource decisions. Similarly, we found that while the officials we spoke with from several DOD offices that manage electronic warfare, including offices within the military services, were unaware of the center's operational status and unclear regarding its mission, roles, and responsibilities, many also thought it to be unlikely that the JEMSCC—as a subordinate center of Strategic Command—would possess the requisite authority to advocate effectively for electronic warfare resource decisions. These concerns were echoed by the independent study, which noted that the center would require strong authorities to substantially influence the allocation of other DOD elements' resources.[25]

Additionally, limited visibility across the department's electronic warfare programs and activities may impede the center's ability to advocate for electronic warfare capabilities development. Specifically, Strategic Command officials told us that they do not have access to information regarding all of the military services' electronic warfare programs and activities, particularly those that are highly classified or otherwise have special access restrictions. In addition, Strategic Command officials told us that they do not have visibility over or participate in rapid acquisitions conducted through the joint capabilities development process. In our March 2012 report on DOD's airborne electronic attack strategy and acquisitions, we reported that certain airborne electronic attack systems in development may offer capabilities that unnecessarily overlap with one another—a condition that appears most prevalent with irregular warfare systems that the services are acquiring under DOD's rapid acquisitions process.[26] The JEMSCC's exclusion from this process is likely to limit its ability to develop the departmentwide perspective necessary for effective advocacy. Moreover, in the absence of clearly defined objectives and an implementation plan outlining major implementation tasks and timelines to measure progress, these potential challenges reduce DOD's level of assurance that the JEMSCC will provide effective departmentwide leadership for electronic warfare capabilities development.

DOD Policy Documents Have Not been Updated to Include All Oversight Roles and Responsibilities for Electronic Warfare

DOD issued two primary directives that provide some guidance for departmentwide oversight of electronic warfare. However, neither of these two directives has been updated to reflect changes in DOD's leadership structures that manage electronic warfare. Federal internal control standards require that organizations establish objectives, clearly define key areas of authority and responsibility, and establish appropriate lines of reporting to aid in the effective and efficient use of resources.[27] Additionally, those standards state that management must continually assess and evaluate its internal control to assure that the actions in place are effective and updated when necessary.

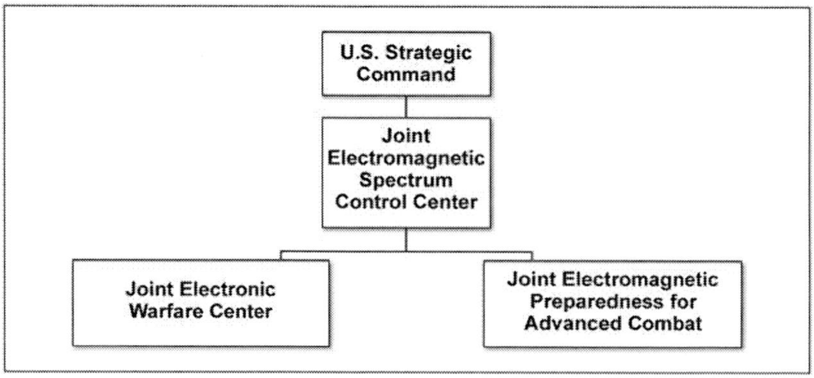

Source: GAO analysis of DOD information.

Figure 4. Joint Electromagnetic Spectrum Control Center under Strategic Command.

DOD's two primary directives that provide some guidance for departmentwide oversight of electronic warfare are:

- DOD Directive 3222.4 (*Electronic Warfare and Command and Control Warfare Countermeasures*)—Designates the Under Secretary of Defense for Acquisition (now Acquisition, Technology, and Logistics) as the focal point for electronic warfare within the department. However, the directive was issued in 1992 and updated in 1994, and does not reflect subsequent changes in policy or organizational structures. For example, the directive does not reflect the establishment of the JEMSCC under Strategic Command.
- DOD Directive 3600.01 (*Information Operations*)—Issued in 2006 and revised in May 2011, this directive provides the department with a framework for oversight of information operations, which was defined as the integrated employment of the core capabilities of electronic warfare, computer network operations, military information support operations (formerly referred to as psychological operations), military deception, and operations to influence, disrupt, corrupt, or usurp adversarial human and automated decision making while protecting that of the United States. However, the definition of oversight responsibilities for information operations has changed, and these changes have not yet been reflected in DOD Directive 3600.01.[28]

DOD Directive 3222.4 has not been updated to reflect the responsibilities for electronic warfare assigned to Strategic Command. Both the December 2008 and April 2011 versions of the *Unified Command Plan* assigned Strategic Command responsibility for advocating for joint electronic warfare capabilities.[29] Similarly, the directive has not been updated to reflect the establishment of the JEMSCC and its associated electronic warfare responsibilities. Specifically, the directive does not acknowledge that JEMSCC has been tasked by Strategic Command as the primary focal point for electronic warfare; rather, the directive designates the Under Secretary of Defense for Acquisition, Technology, and Logistics as the focal point for electronic warfare within DOD. As a result, it is unclear what JEMSCC's roles and responsibilities are in relation to those of the Under Secretary of Defense for Acquisition, Technology, and Logistics. For example, it's unclear what JEMSCC's role will be regarding development of future iterations of the DOD's electronic warfare strategy report to Congress, which is currently produced by the Office of the Under Secretary of Defense for Acquisition, Technology, and Logistics. Also it is unclear what role, if any, the JEMSCC will have in prioritizing electronic warfare investments. Moreover, the directive has not been updated to reflect the Secretary of Defense's memorandum issued in January 2011, which assigned individual capability responsibility for electronic warfare and computer network operations to Strategic Command.

DOD Directive 3600.01 provides both the Under Secretary of Defense for Acquisition, Technology, and Logistics and the Under Secretary of Defense for Intelligence with responsibilities that aid in the oversight of electronic warfare within DOD. However, pursuant to the Defense Secretary's January 2011 memo, the directive is under revision to accommodate changes in roles and responsibilities. Under the current version of DOD Directive 3600.01, the Under Secretary of Defense for Intelligence is charged with the role of Principal Staff Advisor to the Secretary of Defense for information operations. The Principal Staff Advisor is responsible for, among other things, the development and oversight of information operations policy and integration activities as well as the coordination, oversight, and assessment of the efforts of DOD components to plan, program, develop, and execute capabilities in support of information operations requirements.[30] Additionally, the current Directive 3600.01 identifies the Under Secretary of Defense for Acquisition, Technology, and Logistics as responsible for establishing specific policies for the development of electronic warfare as a core capability of information operations.

Under the requirements of DOD acquisition policy, the Under Secretary of Defense for Acquisition, Technology, and Logistics regularly collects cost, schedule, and performance data for major programs.[31] In some cases, the cost information of electronic warfare systems are reported as distinct programs, while in other cases, some electronic warfare systems are subcomponents of larger programs, and cost information is not regularly collected for these separate subsystems. Additionally, the Under Secretary—in coordination with the Army, the Navy, and the Air Force—is developing an implementation road map for electronic warfare science and technology. The road map is supposed to coordinate investments across DOD to accelerate the development and delivery of capabilities. The road map is expected to be completed in late summer of 2012.

The Secretary of Defense issued a memorandum in January 2011 that prompted DOD officials to begin revising DOD Directive 3600.01. The memorandum redefined information operations as "the integrated employment, during military operations, of information-related capabilities in concert with other lines of operation to influence, disrupt, corrupt, or usurp the decision-making of adversaries and potential adversaries while protecting our own." Previously, DOD defined information operations as the "integrated employment of the core capabilities of electronic warfare, computer network operations, psychological operations, military deception, and operations security, in concert with specified supporting and related capabilities, to influence, disrupt, corrupt, or usurp adversarial human and automated decision making while protecting our own." According to DOD officials, the revised definition removed the term core capabilities because it put too much emphasis on the individual core capabilities and too little emphasis on the integration of these capabilities.

Additionally, the memorandum noted that the Under Secretary of Defense for Policy began serving as the Principal Staff Advisor for information operations as of October 1, 2010, and charged the Under Secretary of Defense for Policy with revising DOD Directive 3600.01 to reflect these responsibilities. According to the memorandum, the Principal Staff Advisor is to serve as the single point of fiscal and program accountability for information operations. However, according to DOD officials, this accountability oversight covers only the integration of information operations-related capabilities and does not cover the formerly defined core capabilities of information operations, including electronic warfare and computer network operations. For example, DOD officials stated that the Principal Staff Advisor for information operations would maintain program accountability where information operations-related capabilities were integrated but would not maintain program accountability for all information-related capabilities. However, the memorandum does not clearly describe the specific responsibilities of the Principal Staff Advisor for information operations.

The Secretary's memorandum directed the Under Secretary of Defense for Policy, together with the Undersecretary of Defense (Comptroller) and Director of Cost Analysis and Program Evaluation, to continue to work to develop standardized budget methodologies for information operations-related capabilities and activities. However, these budget methodologies would capture only data related to information operations. For example, according to Under Secretary of Defense for Policy officials, they do not collect or review electronic warfare financial data, but may review this data in the future to determine if it relates to integrated information operations efforts. Officials from the Office of the Under Secretary of Defense for Policy stated that DOD Directive 3600.01 was under revision to reflect these and other changes as directed by the Secretary's memorandum. Until the underlying directive is revised, there may be uncertainty regarding which office has the authority to manage and oversee which programs. Moreover, until this directive is updated, it is not clear where the boundaries are for oversight of electronic warfare between the Under Secretary of Defense for Policy and the Under Secretary of Defense for Acquisition, Technology, and Logistics.

Table 1 compares the oversight roles and responsibilities for electronic warfare as described in the two DOD directives and the Secretary's 2011 policy memorandum.

DOD May Face Challenges in Its Oversight of the Evolving Relationship of Electronic Warfare and Cyberspace Operations

DOD may face challenges in its oversight of electronic warfare because of the evolving relationship between electronic warfare and cyberspace operations, specifically computer network operations; both are information operations-related capabilities.

According to DOD, to ensure all aspects of electronic warfare can be developed and integrated to achieve electromagnetic spectrum control, electronic warfare must be clearly and distinctly defined in its relationship to information operations (to include computer network operations) and the emerging domain of cyberspace. In the previous section, we noted that DOD's directives do not clearly define the roles and responsibilities for the oversight of electronic warfare in relation to the roles and responsibilities for information operations. The current DOD Directive 3600.01 does not clearly specify what responsibilities the Principal Staff Advisor has regarding the integration of information operations-related capabilities—specifically the integration of electronic warfare capabilities with computer network operations.[32]

Further, DOD's fiscal year 2011 electronic warfare strategy report to Congress, which delineated its electronic warfare strategy, stated that the strategy has two, often co-dependent capabilities: traditional electronic warfare and computer network attack, which is part of cyberspace operations. Moreover, according to DOD officials, the relationship between electronic warfare and cyberspace operations—including computer network attack—is still evolving, which is creating both new opportunities and challenges.

Table 1. Department of Defense Electronic Warfare Responsibilities

Organizations assigned electronic warfare responsibilities[a]	**DOD Directive 3222.4, *Electronic Warfare*, July 31, 1992 (Incorporating Change 2, January 28, 1994)**	**DOD Directive 3600.01, *Information Operations*, August 14, 2006 Incorporating Change 1, May 23, 2011)** (Electronic warfare defined as one of the core capabilities of information operations)	**Secretary of Defense, *Memorandum: Strategic Communication and Information Operations in the DOD*, January 25, 2011** (Electronic warfare defined as an information operations-related capability)
Office of the Under Secretary of Defense for Policy	• No responsibility assigned.	• Provide oversight of information operations planning, execution, and related policy guidance, including the establishment of an Office of the Secretary of Defense review process to assess information operations plans and programs submitted by combatant commanders.	• Assigned the Principal Staff Advisor function and respons-ibility for information operations oversight and management. • Tasked to revise DOD Directive 3600.01 and DOD Directive 5111.1, and other relevant policy and doctrine documents to reflect a new definition of information operations. • Assigned as the single point of fiscal and program accountability for information operations.
Office of the Under	• Focal point for electronic warfare within DOD. • Provide guidance on electronic warfare policy.	Establish specific policies for the development and integration of electronic warfare.	• No responsibility assigned..

Table 1. (Continued)

Organizations assigned electronic warfare responsibilities[a]	DOD Directive 3222.4, *Electronic Warfare*, July 31, 1992 (Incorporating Change 2, January 28, 1994)	DOD Directive 3600.01, *Information Operations*, August 14, 2006 Incorporating Change 1, May 23, 2011) (Electronic warfare defined as one of the core capabilities of information operations)	Secretary of Defense, *Memorandum: Strategic Communication and Information Operations in the DOD*, January 25, 2011 (Electronic warfare defined as an information operations-related capability)
Secretary of Defense for Acquisition, Technology, and Logistics	• Provide oversight for development and acquisition of tactical land, sea, air, space, or undersea electronic warfare systems. • Review electronic warfare programs for duplication and maximum multi-service applications. • Provide matrix electronic warfare technical and/or management support within the Office of the Secretary of Defense on request. • Ensure that adequate science and technology programs exist for development and acquisition of electronic warfare systems.	• Develop and maintain a technology investment strategy to support the development, acquisition, and integration of electronic warfare capabilities. Invest in and develop the science and technologies needed to support information operations capabilities.	

Organizations assigned electronic warfare responsibilities[a]	**DOD Directive 3222.4, *Electronic Warfare*, July 31, 1992 (Incorporating Change 2, January 28, 1994)**	**DOD Directive 3600.01, *Information Operations*, August 14, 2006 Incorporating Change 1, May 23, 2011)** (Electronic warfare defined as one of the core capabilities of information operations)	**Secretary of Defense, *Memorandum: Strategic Communication and Information Operations in the DOD*, January 25, 2011** (Electronic warfare defined as an information operations-related capability)
Office of the Under Secretary of Defense for Intelligence	• No responsibility assigned.	• Serve as the Principal Staff Advisor for information operations. • Develop and oversee DOD information operations policy and integration activities. Coordinate, oversee, and assess the efforts of the DOD components to plan, program, develop, and execute capabil-ities in support of information operations requirements. • Establish specific policies for the development and integration of computer network operations, military deception and operational security.	• Relieved of role as the Principal Staff Advisor function and responsibility for information operations oversight and management.

Table 1. (Continued)

Organizations assigned electronic warfare responsibilities[a]	DOD Directive 3222.4, *Electronic Warfare*, July 31, 1992 (Incorporating Change 2, January 28, 1994)	DOD Directive 3600.01, *Information Operations*, August 14, 2006 Incorporating Change 1, May 23, 2011) (Electronic warfare defined as one of the core capabilities of information operations)	Secretary of Defense, *Memorandum: Strategic Communication and Information Operations in the DOD*, January 25, 2011 (Electronic warfare defined as an information operations-related capability)
Strategic Command[b]	• NA[c]	• Shall integrate and coordinate DOD information operations core capabilities (including electronic warfare and computer network operations) that cross geographic areas of responsibility or across the core information operations areas.	• Assigned individual capability responsibilities for electronic warfare and computer network operations.

Source: GAO analysis of DOD documents.

Note: This table presents only those responsibilities that pertain to electronic warfare management and oversight within DOD, including those that relate to electronic warfare as a core or related capability of information operations. The table excludes, for example, responsibilities related to electronic warfare training and intergovernmental coordination, and the entities assigned such responsibilities.

[a] These organizations may be assigned non-electronic warfare capabilities in the listed documents. Here we only include those responsibilities that are related to electronic warfare, including electronic warfare as a core capability or related capability of information operations.

[b] Figure only provides Strategic Command responsibilities unique from those assigned to all combatant commands.

[c] DOD Directive 3222.4 predates the creation of Strategic Command in 2002.

There will be operations and capabilities that blur the lines between cyberspace operations and electronic warfare because of the continued expansion of wireless networking and the integration of computers and radio frequency communications. According to cognizant DOD officials, electronic warfare capabilities may permit use of the electromagnetic spectrum as a maneuver space for cyberspace operations. For example, electronic warfare capabilities may serve as a means of accessing otherwise inaccessible networks to conduct cyberspace operations; presenting new opportunities for offensive action as well as the need for defensive preparations.

Current DOD doctrine partially describes the relationship between electronic warfare and cyberspace operations. Specifically, current joint doctrine for electronic warfare, which was last updated in February 2012, states that since cyberspace requires both wired and wireless links to transport information, both offensive and defensive cyberspace operations may require use of the electromagnetic spectrum for the enabling of effects in cyberspace. Due to the complementary nature and potential synergistic effects of electronic warfare and computer network operations, they must be coordinated to ensure they are applied to maximize effectiveness.[33] When wired access to a computer system is limited, electromagnetic access may be able to successfully penetrate the computer system. For example, use of an airborne weapons system to deliver malicious code into cyberspace via a wireless connection would be characterized as "electronic warfare-delivered computer network attack." In addition, the doctrine mentions that electronic warfare applications in support of homeland defense are critical to deter, detect, prevent, and defeat external threats such as cyberspace threats.

DOD has not yet published specific joint doctrine for cyberspace operations, as we previously reported.[34] We recommended, among other things, that DOD establish a time frame for deciding whether to proceed with a dedicated joint doctrine publication on cyberspace operations and update existing cyber-related joint doctrine.[35] DOD agreed and has drafted, but not yet issued, the joint doctrine for cyberspace operations. According to U.S. Cyber Command officials, it is unclear when the doctrine for cyberspace operations will be issued.

The military services also have recognized the evolving relationship between electronic warfare and cyberspace operations. For example, to address future challenges, the U.S. Army Training and Doctrine Command conducted an assessment on how the Army's future force will leverage cyberspace operations and found that the Army's current vocabulary—including terms such as computer network operations, electronic warfare, and information operations—will become increasingly inadequate. According to the Army, these terms are becoming outdated as the operational environment rapidly changes due to factors such as technologic convergence of computer and telecommunication networks, astonishing rates of technologic advancements, and the global proliferation of information and communications technology. According to a Navy official, the Navy recognizes the evolving relationship between electronic warfare and cyberspace operations and is moving toward defining that relationship. However, the Navy first is working to define the relationship between electronic warfare and electromagnetic spectrum operations. In addition, Air Force Instruction 10-706, *Electronic Warfare Operations*,[36] states that traditional electronic warfare capabilities are beginning to overlap with cyberspace areas, which is resulting in an increased number of emerging targets such as non-military leadership networks and positioning, navigation, and timing networks.

According to U.S. Cyber Command officials, it is important to understand how electronic warfare and cyberspace operations capabilities might be used in an operational setting. Such information could then inform the further development of doctrine. U.S. Cyber Command officials stated that they have participated in regular meetings with representatives from the military services, the National Security Agency, defense research laboratories, and others, to discuss the relationship of electronic warfare and cyberspace operations. Moreover, the Under Secretary for Acquisition, Technology, and Logistics, has established steering committees that are developing road maps for the Secretary of Defense's seven designated science and technology priority areas—one of which is cyberspace operations and another is electronic warfare.

CONCLUSION

DOD faces significant challenges in operating in an increasingly complex electromagnetic environment. Therefore, it is important that DOD develop a comprehensive strategy to ensure departmental components are able to integrate electronic warfare capabilities into all phases of military operations and maintain electromagnetic spectrum access and maneuverability. DOD would benefit from a strategy that includes implementing parties, roles, responsibilities, and performance measures, which can help ensure that entities are effectively supporting such objectives, and linking resources and investments to key activities necessary to meet strategic goals and priorities. In the absence of a strategy that fully addresses these and other key elements, the DOD components and military services responsible for implementing this strategy, evaluating progress, and ensuring accountability may lack the guidance necessary to prioritize their activities and establish milestones that are necessary to achieve intended results within a reasonable time frame. Moreover, as a result, DOD may not be effectively managing its electronic warfare programs and activities or using its resources efficiently. For example, an effective strategy could help DOD reduce the potential for unnecessary overlap in the airborne electronic attack acquisition activities on which we have previously reported.

The military's increasing reliance on the electromagnetic spectrum—coupled with a fiscally constrained environment and critical gaps in electronic warfare management—highlights the need for an effective governance framework for managing and conducting oversight of the department's electronic warfare activities. The absence of such a framework can exacerbate management challenges, including those related to developing and implementing an effective strategy and coordinating activities among stakeholders. Without additional steps to define the purpose and activities of the JEMSCC, DOD lacks reasonable assurance that this center will provide effective departmentwide leadership for electronic warfare capabilities development and ensure the effective and efficient use of its resources. As we previously reported, DOD acknowledges a leadership void that makes it difficult to ascertain whether the current level of investment is optimally matched with the existing capability gaps. Leveraging resources and acquisition efforts across DOD—not just by sharing information, but through shared partnerships and investments—can simplify developmental efforts, improve interoperability among systems and combat forces, and could decrease future operating and support costs. Such successful outcomes can position the department to maximize the returns it gets on its electronic warfare investments. In addition, multiple organizations are involved with electronic warfare and outdated guidance regarding management and oversight may limit the effectiveness of their activities. Both the Under Secretary of Defense for Acquisition, Technology, and Logistics and the JEMSCC have been identified as the focal point for electronic warfare within the department, yet it is unclear what each organization's roles and responsibilities are in relation to one another. Further, each organization's management responsibilities related to future iterations of the electronic warfare strategy report to Congress and working with the military services to prioritize investments remain unclear. Updating electronic warfare directives and policy documents to clearly define oversight roles and responsibilities for electronic warfare—including any roles and responsibilities related to managing the relationship between electronic warfare and information operations or electronic warfare and cyberspace operations, specifically computer network operations—would help ensure that all aspects of electronic warfare can be developed and integrated to achieve electromagnetic spectrum control.

RECOMMENDATIONS FOR EXECUTIVE ACTION

To improve DOD's management, oversight, and coordination of electronic warfare policy and programs, we recommend that the Secretary of Defense take the following three actions:

- Direct the Under Secretary of Defense for Acquisition, Technology, and Logistics, in coordination with the Under Secretary of Defense for Policy and Strategic Command, and others, as appropriate, to include at a minimum the following information in the fiscal years 2013 through 2015 strategy reports for electronic warfare:
    - Performance measures to guide implementation of the strategy and help ensure accountability. These could include milestones to track progress toward closing the 34 capability gaps identified by DOD studies.
    - Resources and investments necessary to implement the strategy, including those related to key activities, such as developing electronic warfare organizational structures and leadership.
    - The parties responsible for implementing the department's strategy, including specific roles and responsibilities.
- Direct the Commander of Strategic Command to define the objectives of the Joint Electromagnetic Spectrum Control Center and issue an implementation plan outlining major implementation tasks and timelines to measure progress.
- Direct the Under Secretary of Defense for Policy, in concert with the Under Secretary of Defense for Acquisition, Technology, and Logistics, as appropriate, to update key departmental guidance regarding electronic warfare—including DOD Directives 3222.4 (*Electronic Warfare and Command and Control Warfare Countermeasures*) and 3600.01 (*Information Operations*)—to clearly

define oversight roles and responsibilities of and coordination among the Under Secretary of Defense for Policy; the Under Secretary of Defense for Acquisition, Technology, and Logistics; and the Joint Electromagnetic Spectrum Control Center. Additionally, the directives should clarify, as appropriate, the oversight roles and responsibilities for the integration of electronic warfare and cyberspace operations, specifically computer network operations.

AGENCY COMMENTS AND OUR EVALUATION

In written comments on a draft of this report, DOD partially concurred with our first recommendation and concurred with our other two recommendations. Regarding our recommendation that DOD include in future strategy reports for electronic warfare, at a minimum, information on (1) performance measures to guide implementation of the strategy, (2) resources and investments necessary to implement the strategy, and (3) parties responsible for implementing the strategy, the department stated that it continues to refine the annual strategy reports for electronic warfare and will expand upon resourcing plans and organization roles; however, the department stated that the strategy was not intended to be prescriptive with performance measures. As we have previously stated, the inclusion of performance measures can aid entities responsible for implementing DOD's electronic warfare strategy in establishing priorities and milestones to aid in achieving intended results within reasonable time frames. We also have noted that performance measures can enable more effective oversight and accountability as progress toward meeting a strategy's goals may be measured, thus helping to ensure the strategy's successful implementation. We therefore continue to believe this recommendation has merit.

DOD concurred with our remaining two recommendations that (1) the Commander of Strategic Command define the objectives of the JEMSCC and issue an implementation plan for the center and (2) DOD update key departmental guidance regarding electronic warfare. These steps, if implemented, will help to clarify the roles and responsibilities of electronic warfare management within the department and aid in the efficient and effective use of resources.

Brian J. Lepore
Director, Defense Capabilities and Management

Appendix I. Scope and Methodology

To assess the extent to which DOD has developed a strategy to manage electronic warfare we evaluated DOD's fiscal year 2011 and 2012 electronic warfare strategy reports to Congress[37] against prior GAO work on strategic planning, that indentified six desirable characteristics of a strategy.[38] The characteristics GAO previously identified are: (1) purpose, scope, and methodology; (2) problem definition and risk assessment; (3) goals, subordinate objectives, activities, and performance measures; (4) resources, investments, and risk management; (5) organizational roles, responsibilities, and coordination; and (6) integration and implementation. While these characteristics were identified in our past work as desirable components of national-level strategies, we determined that they also are relevant to strategies of varying scopes, including defense strategies involving complex issues. For example, identifying organizational roles, responsibilities, and coordination mechanisms is key to allocating authority and responsibility for implementing a strategy. Further, goals, objectives, and performance measures provide concrete guidance for implementing a strategy, allowing implementing parties to establish priorities and milestones, and providing them with the flexibility necessary to pursue and achieve those results within a reasonable time frame. Full descriptions of these characteristics are contained in appendix II.

We determined that the strategy "addressed" a characteristic when it explicitly cited all elements of a characteristic, even if it lacked specificity and details and could thus be improved upon.

The strategy "partially addressed" a characteristic when it explicitly cited some, but not all, elements of a characteristic. Within our designation of "partially addressed," there may be wide variation between a characteristic for which most of the elements were addressed and a characteristic for which few of the elements of the characteristic were addressed.

The strategy "did not address" a characteristic when it did not explicitly cite or discuss any elements of a characteristic, and/or any implicit references were either too vague or general. To supplement this analysis and gain further insight into issues of strategic import, we also reviewed other relevant strategic planning documents—such DOD's *National Defense Strategy*,[39] *Strategic Spectrum Plan*,[40] and *Net-Centric Spectrum Management Strategy*[41]—and interviewed cognizant officials from organizations across the department, including the Office of the Under Secretary of Defense for Acquisition, Technology, and Logistics; U.S. Strategic Command; and the Joint Chiefs of Staff.

To determine the extent to which DOD has planned, organized, and implemented an effective governance structure to oversee its electronic warfare policy and programs and their relationship with cyberspace operations, we reviewed and analyzed relevant DOD policies, doctrine, plans, briefings, and studies.

Specifically, to determine how DOD has allocated electronic warfare authorities and responsibilities across the department, we reviewed and analyzed DOD policy, including DOD Directive 3222.4, *Electronic Warfare and Command and Control Warfare Countermeasures*;[42] DOD Directive 3600.01, *Information Operations*;[43] and the Secretary of Defense's *Memorandum: Strategic Communication and Information Operations in the DOD*.[44] We also reviewed relevant joint doctrine publications, such as Joint Publications 3-13, *Information Operations*[45] and 3-13.1, *Electronic Warfare*;[46] plans, including the 2008 and 2011 *Unified Command Plans*;[47] strategic documents, such as DOD's fiscal year 2011 and 2012 electronic warfare strategy reports to Congress;[48] and classified and unclassified briefings, and studies related to DOD's identification of and efforts to address electronic warfare capability gaps, including DOD's 2009 *Electronic Warfare Initial Capabilities Document*.[49] We also reviewed DOD and military service reports, plans, concepts of operation, and outside studies that discuss DOD's definitions of electronic warfare and cyberspace operations. In addition, we interviewed cognizant DOD officials to obtain information and perspectives regarding policy, management, and technical issues related to electronic warfare, information operations, electromagnetic spectrum control, and cyberspace operations.

In addressing both of our objectives, we obtained relevant documentation from and/or interviewed officials from the following DOD offices, combatant commands, military services, and combat support agencies:

- Office of the Under Secretary of Defense for Policy
- Office of the Under Secretary of Defense for Intelligence
- Office of the Under Secretary of Defense for Acquisition, Technology, and Logistics
- Office of the Assistant Secretary of Defense for Networks and Integration/DOD Chief Information Officer
- Joint Chiefs of Staff
- Combatant Commands
    - U.S. Cyber Command, Fort Meade, Maryland
    - U.S. Pacific Command, Camp H.M. Smith, Hawaii

    - U.S. Strategic Command, Offutt Air Force Base, Nebraska
        - Joint Electromagnetic Spectrum Control Center, Offutt Air Force Base, Nebraska
        - Joint Electronic Warfare Center, Lackland Air Force Base, Texas
- U.S. Army
    - Office of the Deputy Chief of Staff of the Army for Operations, Plans, and Training, Electronic Warfare Division
    - Training and Doctrine Command, Combined Arms Center Electronic Warfare Proponent Office, Fort Leavenworth, Kansas
- U.S. Air Force—Electronic Warfare Division
- U.S. Marines Corps—Headquarters, Electronic Warfare Branch
- U.S. Navy
    - Office of the Deputy Chief of Naval Operations for Information Dominance Electronic and Cyber Warfare Division
    - Naval Sea Systems Command, Naval Surface Warfare Center, Crane, Indiana
    - Naval Sea Systems Command, Program Executive Office for Integrated Warfare Systems
    - Navy Fleet Forces Cyber Command, Fleet Electronic Warfare Center, Joint Expeditionary Base Little Creek-Fort Story , Virginia
- Combat Support Agencies
    - Defense Information Systems Agency—Defense Spectrum Organization
    - National Security Agency, Fort Meade, Maryland

We conducted this performance audit from July 2011 to July 2012 in accordance with generally accepted government auditing standards. Those standards require that we plan and perform the audit to obtain sufficient, appropriate evidence to provide a reasonable basis for our findings and conclusions based on our audit objectives.

We believe that the evidence obtained provides a reasonable basis for our findings and conclusions based on our audit objectives.

APPENDIX II. DESIRABLE STRATEGY CHARACTERISTICS

We previously identified a set of desirable strategy characteristics to aid responsible parties in implementation, enhance the strategies' usefulness in resource and policy decisions, and to better ensure accountability.[50] Table 2 provides a brief description of each characteristic and its benefit.

Table 2. Summary of Desirable Characteristics for a Strategy, Their Description, and Benefit

Characteristic	Summary description	Benefit
Purpose, scope, and methodology	Addresses why the strategy was produced, the scope of its coverage, and the process by which it was developed.	A complete description of the purpose, scope, and methodology in a strategy could make the document more useful to the entities it is intended to guide, as well as to oversight organizations, such as Congress.
Problem definition and risk assessment	Addresses the particular problems and threats the strategy is directed toward.	Use of common definitions promotes more effective intergovernmental operations and more accurate monitoring of expenditures, thereby eliminating problematic concerns. Comprehensive assessments of vulnerabilities, including risk assessments, can help identify key factors external to an organization that can significantly affect that organization's attainment of its goals and objectives and can help identify risk potential if such problem areas are not effectively addressed.
Goals, subordinate objectives, activities, and performance measures	Addresses what the strategy is trying to achieve, steps to achieve those results, as well as the priorities, milestones, and performance measures to gauge results.	Better identification of priorities, milestones, and performance measures can aid implementing entities in achieving results in specific time frames—and could enable more effective oversight and accountability.

Table 2. (Continued)

Characteristic	Summary description	Benefit
Resources, investments, and risk management	Addresses what the strategy will cost, the sources and types of resources and investments needed, and where resources and investments should be targeted based on balancing risk reductions with costs.	Guidance on resource, investment, and risk management could help implementing entities allocate resources and investments according to priorities and constraints, track costs and performance, and shift such investments and resources as appropriate. Such guidance could also assist organizations in developing more effective programs to stimulate desired investments and leverage finite resources.
Organizational roles, responsibilities, and coordination	Addresses who will be implementing the strategy, what their roles will be compared to others, and mechanisms for them to coordinate their efforts.	Inclusion of this characteristic in a strategy could be useful to organizations and other stakeholders in fostering coordination and clarifying specific roles, particularly where there is overlap, and thus enhancing both implementation and accountability.
Integration and implementation	Addresses how a strategy relates to other strategies' goals, objectives and activities, and to subordinate levels of government and their plans to implement the strategy.	Information on this characteristic in a strategy could build on the aforementioned organizational roles and responsibilities—and thus further clarify the relationships between various implementing entities. This, in turn, could foster effective implementation and accountability.

Source: GAO.

End Notes

[1] GAO, *Airborne Electronic Attack: Achieving Mission Objectives Depends on Overcoming Acquisition Challenges*, GAO-12-175 (Washington, D.C.: Mar. 29, 2012).

[2] Potential near-peer adversaries include countries capable of waging large-scale conventional war on the United States. These nation-states are characterized as having nearly comparable diplomatic, informational, military, and economic capacity to the United States.

[3] DOD defines cyberspace operations, which includes computer network operations, as the employment of cyberspace capabilities where the primary purpose is to achieve military objectives or effects in or through cyberspace. DOD documents that discuss the relationship between electronic warfare and cyberspace operations use several different cyber-related terms, including cyberspace, cyber operations, computer network operations, and computer network attack. In addition, according to DOD, the definition of information operations includes the term computer network operations because it is an information operations-related capability. To provide clarity in this report, we generally use the term cyberspace operations in our discussion of the relationship between electronic warfare and cyberspace operations and computer network operations in our discussions concerning information operations-related capabilities.

[4] Center for Strategic and International Studies, *Organizing for Electro-Magnetic Spectrum Control* (Washington, D.C.: May 2010).

[5] GAO . For additional information, see GAO, *2012 Annual Report: Opportunities to Reduce Duplication, Overlap and Fragmentation, Achieve Savings, and Enhance Revenue*, GAO-12-342SP (Washington, D.C.: Feb. 28, 2012).

[6] GAO .

[7] According to DOD, electromagnetic spectrum control is freedom of action in the electromagnetic operational environment, which is achieved through the coordinated implementation of joint electromagnetic spectrum operations, which includes electronic warfare, with other lethal and nonlethal operations impacting the electromagnetic operational environment. See Joint Chiefs of Staff, Joint Publication 3-13.1, *Electronic Warfare* (Feb. 8, 2012).

[8] Pub. L. No. 111-84, § 1053 (a).

[9] Pub. L. No. 111-84, § 1053 (b) delineates several other requirements for the strategy, including a list of electronic warfare acquisition programs and research and development projects with associated program or project information.

[10] Office of the Under Secretary of Defense for Acquisition, Technology, and Logistics, Report to the Congressional Defense Committees. *Electronic Warfare Strategy of the Department of Defense* (Washington, D.C.: October 2010). The fiscal year 2012 report is classified. As of July 2012, the fiscal year 2013 electronic warfare strategy report was still being drafted, according to officials from the Office of the Under Secretary of Defense for Acquisition, Technology, and Logistics.

[11] See GAO, *Combating Terrorism: Evaluation of Selected Characteristics in National Strategies Related to Terrorism*, GAO-04-408T (Washington, D.C.: Feb. 3, 2004). This testimony identified six characteristics of an effective strategy. While these characteristics were identified in our past work as desirable components of national-level strategies, we determined that they also are relevant to strategies of varying scopes, including defense strategies involving complex issues. For example, identifying organizational roles, responsibilities and coordination mechanisms is key to allocating authority and responsibility for implementing a strategy. Further, goals, objectives, and performance measures provide concrete guidance for implementing a strategy, allowing implementing parties to establish priorities and milestones, and providing them with the flexibility necessary to pursue and achieve those results within a reasonable timeframe. Full descriptions of these characteristics are contained in appendix II. See also GAO, *Influenza Pandemic: DOD Has Taken Important Actions to Prepare, but Accountability, Funding, and Communications Need to be Clearer and Focused Departmentwide*, GAO-06-1042 (Washington, D.C.: Sept. 21, 2006) ; GAO, *Defense Space Activities: National Security Space Strategy Needed to Guide Future DOD Space Efforts*, GAO-08-431R (Washington, D.C.: Mar. 27, 2008); and GAO, *Intelligence, Surveillance, and Reconnaissance: DOD*

Needs a Strategic, Risk-Based Approach to Enhance Its Maritime Domain Awareness, GAO-11-621 (Washington, D.C.: June 20, 2011).

[12] GAO .

[13] According to DOD officials, the electronic warfare integrated planning team was established by the Office of the Under Secretary of Defense for Acquisition, Technology and Logistics to develop guidelines for electronic warfare investment strategy.

[14] By "partnering," we refer to shared, or joint, responsibilities among implementing parties where there is otherwise no clear or established hierarchy of lead and support functions.

[15] See Department of Defense, *Strategic Spectrum Plan* (February 2008); and Assistant Secretary of Defense (Networks and Information Integration) Department of Defense Chief Information Officer, *Department of Defense Net-Centric Spectrum Management Strategy (*Aug. 3, 2006).

[16] The Joint Requirements Oversight Council, among other things, assists the Chairman of the Joint Chiefs of Staff in (1) identifying, assessing, establishing priority levels for, and validating joint military requirements, including existing systems and equipment, to meet the *National Military Strategy*; (2) considering trade-offs among cost, schedule, and performance objectives for joint military requirements; and (3) reviewing the estimated level of resources required to fulfill each joint military requirement, and establishing an objective for the overall period of time within which an initial operational capability should be delivered to meet each joint military requirement.

[17] Specifically, in October 2009, the Joint Requirements Oversight Council tasked Strategic Command and the now disestablished U.S. Joint Forces Command to assess both the technical issues related to the electronic warfare problem and the organizational structure and management approach required to respond to emerging electromagnetic spectrum threats. In response, Strategic Command and U.S. Joint Forces Command produced a classified report providing potential organizational solutions. As part of this process, the Center for Strategic and International Studies conducted its review as an independent analysis that also provided organizational alternatives. See Center for Strategic and International Studies, *Organizing for Electro-Magnetic Spectrum Control* (Washington, D.C.: May 2010).

[18] See DOD, *Unified Command Plan* (Washington, D.C.: Dec. 17, 2008), which establishes the missions, responsibilities, and geographic areas of responsibility among the combatant commanders. This plan also tasks the Commander of Strategic Command with advocating for electronic warfare capabilities.

[19] GAO, *Standards for Internal Control in the Federal Government*, GAO/AIMD-00-21.3.1 (Washington, D.C.: Nov. 1999).

[20] For example, see GAO, *Reserve Forces: Army Needs to Finalize an Implementation Plan and Funding Strategy for Sustaining an Operational Reserve Force*, GAO-09-898 (Washington, D.C.: Sept. 17, 2009).

[21] The Joint Electronic Warfare Center, prior to incorporation in the JEMSCC, was tasked with integrating current and emerging joint electronic warfare effects to ensure spectrum control for global military operations. The center provided near-term operational solutions and advocated for long-term electromagnetic capabilities. The Joint Electromagnetic Preparedness for Advanced Combat organization was charged with advancing and improving joint warfighter effectiveness and combat capability by conducting vulnerability assessments of electromagnetic spectrum dependent capabilities, architectures, technologies, and tactics, techniques, and procedures.

[22] Kinetic capabilities focus on destroying forces through the application of physical effects.

[23] In 2003, DOD created the Joint Capabilities Integration and Development System to guide the development of capabilities from a joint perspective. This system was established to provide the department with an integrated, collaborative process to identify and guide development of new capabilities that address the current and emerging security environment. DOD's Joint Requirements Oversight Council oversees the joint capabilities development process

and participates in the development of joint requirements, which includes the identification and analysis and synthesis of capability gaps and the council's subsequent validation of capability needs. The council makes recommendations to the Chairman of the Joint Chiefs of Staff, who advises the Secretary of Defense about what capabilities to invest in as part of DOD's budget process. Before making investment decisions, the military services consider the validated capabilities during their planning, programming, and budgeting processes and make decisions among competing investments.

[24] See GAO, *Defense Management: Perspectives on the Involvement of the Combatant Commands in the Development of Joint Requirements*, GAO-11-527R (Washington, D.C.: May 20, 2011).

[25] Center for Strategic and International Studies, *Organizing for Electro-Magnetic Spectrum Control* (Washington, D.C.: May 2010).

[26] GAO and GAO-12-342SP .

[27] GAO .

[28] Secretary of Defense, *Memorandum: Strategic Communication and Information Operations in the DOD* (Washington, D.C.: Jan. 25, 2011).

[29] Department of Defense, *Unified Command Plan* (Washington, D.C.: Dec. 17, 2008) and Department of Defense, *Unified Command Plan* (Washington, D.C.: Apr. 6, 2011).

[30] DOD Directive 3600.01 refers to the Principal Staff Assistant position while the Secretary's January 2011 memorandum refers to the Principal Staff Advisor position. According to DOD officials, these terms refer to the same position. We use the term Principal Staff Advisor in this report.

[31] See DOD Directive 5000.01 *The Defense Acquisition System* (Washington, D.C.: certified current as of Nov. 20, 2007) and DOD Instruction 5000.02 *Operation of the Defense Acquisition System* (Washington, D.C.: Dec. 8, 2008).

[32] DOD Directive 5143.01 *Under Secretary of Defense for Intelligence* (Washington, D.C.: Nov. 23, 2005) provides the responsibilities of the Under Secretary and restates the Principal Staff Advisor duties stated in DOD Directive 3600.01.

[33] Joint Publication 3-13.1.

[34] See GAO, *Defense Department Cyber Efforts, DOD Faces Challenges in Its Cyber Activities*, GAO-11-75 (Washington, D.C.: July 25, 2011).

[35] GAO .

[36] Air Force Instruction, 10-706, *Electronic Warfare Operations* (Washington, D.C.: Nov. 3, 2010).

[37] Office of the Under Secretary of Defense for Acquisition, Technology, and Logistics, Report to the Congressional Defense Committees. *Electronic Warfare Strategy of the Department of Defense* (Washington, D.C.: October 2010). The fiscal year 2012 report is classified.

[38] GAO.

[39] Department of Defense, *National Defense Strategy* (Washington, D.C.: June 2008).

[40] Department of Defense, *Strategic Spectrum Plan* (Washington, D.C.: February 2008).

[41] Department of Defense Chief Information Officer, *Department of Defense Net-Centric Spectrum Management Strategy* (Washington, D.C.: Aug. 3, 2006).

[42] DOD Directive 3222.4, *Electronic Warfare and Command and Control Warfare Countermeasures* (Washington, D.C.: July 31, 1992, Incorporating Change 2, Jan. 28, 1994).

[43] DOD Directive 3600.01, *Information Operations* (Washington, D.C.: Aug. 14, 2006, Incorporating Change 1, May 23, 2011).

[44] Secretary of Defense, *Memorandum: Strategic Communication and Information Operations in the DOD* (Washington, D.C.: Jan. 25, 2011).

[45] Chairman, Joint Chiefs of Staff, Joint Publication 3-13, *Information Operations* (Washington, D.C.: Feb. 13, 2006).

[46] Chairman, Joint Chiefs of Staff, Joint Publication 3-13.1, *Electronic Warfare* (Washington, D.C.: Feb. 8, 2012).

[47] Department of Defense, *Unified Command Plan* (Washington, D.C.: Dec. 17, 2008) and Department of Defense, *Unified Command Plan* (Washington, D.C.: Apr. 6, 2011).

[48] Office of the Under Secretary of Defense for Acquisition, Technology, and Logistics, Report to the Congressional Defense Committees. *Electronic Warfare Strategy of the Department of Defense* (Washington, D.C.: October 2010). The fiscal year 2012 report is classified.

[49] Department of Defense, *Electronic Warfare Initial Capabilities Document Unclassified Extract* (Washington, D.C.: Sept. 22, 2009). The full version of this report is classified.

[50] GAO, *Combating Terrorism: Evaluation of Selected Characteristics in National Strategies Related to Terrorism*, GAO-04-408T (Washington, D.C.: Feb. 3, 2004).

In: Electronic Warfare
Editors: M. Boyd and P.T. Chapman
ISBN: 978-1-62257-792-7

Chapter 2

AIRBORNE ELECTRONIC ATTACK: ACHIEVING MISSION OBJECTIVES DEPENDS ON OVERCOMING ACQUISITION CHALLENGES*

United States Government Accountability Office

WHY GAO DID THIS STUDY

Airborne electronic attack involves the use of aircraft to neutralize, destroy, or suppress enemy air defense and communications systems. Proliferation of sophisticated air defenses and advanced commercial electronic devices has contributed to the accelerated appearance of new weapons designed to counter U.S. airborne electronic attack capabilities. GAO was asked to assess (1) the Department of Defense's (DOD) strategy for acquiring airborne electronic attack capabilities, (2) progress made in developing and fielding systems to meet airborne electronic attack mission requirements, and (3) additional actions taken to address capability gaps. To do this, GAO analyzed documents related to mission requirements, acquisition and budget needs, development plans, and performance, and interviewed DOD officials.

---

* This is an edited, reformatted and augmented version of the Highlights of GAO-12-175, a report to the Committee on Armed Services, House of Representatives, dated March 2012.

WHAT GAO RECOMMENDS

GAO recommends that DOD conduct program reviews for certain new, key systems to assess cost, schedule, and performance; determine the extent to which the most pressing capability gaps can be met and take steps to fill them; align service investments in science and technology with the departmentwide electronic warfare priority; and review capabilities provided by certain planned and existing systems to ensure investments do not overlap. DOD agreed with three recommendations and partially agreed with the two aimed at reducing potential overlap among systems. DOD plans to assess coordination among systems, whereas GAO sees opportunities for consolidation, as discussed in the report.

WHAT GAO FOUND

The Department of Defense's (DOD) evolving strategy for meeting airborne electronic attack requirements centers on acquiring a family of systems, including traditional fixed wing aircraft, low observable aircraft, unmanned aerial systems, and related mission systems and weapons. DOD analyses dating back a decade have identified capability gaps and provided a basis for service investments, but budget realities and lessons learned from operations in Iraq and Afghanistan have driven changes in strategic direction and program content. Most notably, DOD canceled some acquisitions, after which the services revised their operating concepts for airborne electronic attack. These decisions saved money, allowing DOD to fund other priorities, but reduced the planned level of synergy among systems during operations. As acquisition plans have evolved, capability limitations and sustainment challenges facing existing systems have grown, prompting the department to invest in system improvements to mitigate shortfalls.

DOD is investing in new airborne electronic attack systems to address its growing mission demands and to counter anticipated future threats. However, progress acquiring these new capabilities has been impeded by developmental and production challenges that have slowed fielding of planned systems. Some programs, such as the Navy's EA-18G Growler and the Air Force's modernized EC-130H Compass Call, are in stable production and have completed significant amounts of testing. Other key programs, like the Navy's Advanced Anti-Radiation Guided Missile, have required additional time and funding to address technical challenges, yet continue to face execution risks. In addition, certain systems in development may offer capabilities that overlap with one another—a situation brought on in part by DOD's fragmented urgent operational needs processes. Although services have shared technical data among these programs, they continue to pursue unique systems intended to counter similar threats. As military operations in Iraq and Afghanistan decrease, opportunities exist to consolidate current acquisition programs across services. However, this consolidation may be hampered by DOD's acknowledged leadership deficiencies within its electronic warfare enterprise, including the lack of a designated, joint entity to coordinate activities. Furthermore, current and planned acquisitions will not fully address materiel-related capability gaps identified by DOD—including some that date back 10 years. Acquisition program shortfalls will exacerbate these gaps.

To supplement its acquisition of new systems, DOD is undertaking other efforts to bridge existing airborne electronic attack capability gaps. In the near term, services are evolving tactics, techniques, and procedures for existing systems to enable them to take on additional mission tasks. These activities maximize the utility of existing systems and better position operators to complete missions with equipment currently available. Longer-term solutions, however, depend on DOD successfully capitalizing on its investments in science and technology. DOD has recently taken actions that begin to address long-standing coordination shortfalls in this area, including designating electronic warfare as a priority investment area and creating a steering council to link capability gaps to research initiatives. These steps do not preclude services from funding their own research priorities ahead of departmentwide priorities. DOD's planned implementation roadmap for electronic warfare offers an opportunity to assess how closely component research investments are aligned with the departmentwide priority.

ABBREVIATIONS

AARGM	Advanced Anti-Radiation Guided Missile
AESA	Active Electronically Scanned Array
ASD (R&E)	Assistant Secretary of Defense for Research and Engineering
CEASAR	Communications Electronic Attack with Surveillance and Reconnaissance
DARPA	Defense Advanced Research Projects Agency

DOD	Department of Defense
HARM	High Speed Anti-Radiation Missile
ICAP	Improved Capability
IDECM	Integrated Defensive Electronic Countermeasures
ITALD	Improved Tactical Air Launched Decoy
J-UCAS	Joint Unmanned Combat Air Systems
LAIRCM	Large Aircraft Infrared Countermeasures
MALD	Miniature Air Launched Decoy
MALD-J	Miniature Air Launched Decoy—Jammer
RDT&E	Research, Development, Testing, and Evaluation
TALD	Tactical Air Launched Decoy

March 29, 2012

The Honorable Howard P. McKeon
Chairman
The Honorable Adam Smith
Ranking Member
Committee on Armed Services
House of Representatives

Airborne electronic attack capabilities are key enablers for U.S. military operations ranging from irregular warfare[1] to major combat against potential near-peer adversaries.[2] Airborne electronic attack involves the use of aircraft to neutralize, destroy, or temporarily degrade (suppress) enemy air defense and communications systems, either through destructive or disruptive means. These aircraft employ a variety of mission systems and weapons to prosecute threats, and they rely on defensive countermeasures to provide additional protection.

Global proliferation of more sophisticated air defenses and advanced, commercial digital electronic devices has contributed to the accelerated appearance of new weapons designed to counter U.S. airborne electronic attack capabilities and limit U.S. access to theaters of combat. These weapons—some held by both nation-state and nonstate actors—vary from advanced, integrated air defense systems to simpler, digital radio frequency memory devices. As the range of adversary weapons increases, electronic jammers and other equipment must respond with improved capabilities or may have to operate farther from the battle, lessening their effectiveness.

In light of these developments, you asked us to review the Department of Defense's (DOD) airborne electronic attack capabilities and investment plans. In response to this request, we assessed (1) the department's strategy for acquiring airborne electronic attack capabilities; (2) progress made developing and fielding systems to meet airborne electronic attack mission requirements; and (3) additional compensating actions taken by the department to address capability gaps, including improvements to tactics, techniques, and procedures and investments in science and technology. In a separate report, we plan to address the effectiveness of the department's governance structure for overseeing its electronic warfare policies and priorities and the relationship between electronic warfare and cyber operations.

To assess the department's strategy for acquiring airborne electronic attack capabilities, we analyzed documents outlining mission requirements and acquisition needs including the 2009 Electronic Warfare Initial Capabilities Document, service roadmaps related to airborne electronic attack, budget documents, and program briefings. We corroborated this information through discussions with officials responsible for managing airborne electronic attack requirements and systems, including the Office of the Under Secretary of Defense for Acquisition, Technology and Logistics; Navy, Air Force, Army, and Marine Corps requirements branches; U.S. Strategic Command; and the Joint Staff. To assess progress made developing and fielding systems to meet airborne electronic attack mission requirements, we analyzed materials outlining acquisition plans, costs, and performance outcomes including, capabilities documents, program schedules, test reports, budget submissions, and program briefings. These same materials afforded information on key attributes of individual airborne electronic attack systems, which we used to assess potential overlap among systems in development. Further, we identified persisting capability gaps by reviewing DOD analyses related to airborne electronic attack requirements. To supplement our analyses and gain additional visibility and perspective into these issues, we conducted numerous interviews with DOD officials charged with managing airborne electronic attack requirements and those responsible for developing, acquiring, and testing airborne electronic attack systems. To assess additional compensating actions taken by the department to address airborne electronic attack capability gaps, we reviewed service documents outlining recent improvements and refinements to tactics, techniques, and procedures for key airborne electronic attack aircraft. We also reviewed broad agency announcements to understand ongoing science and technology activities. We corroborated this information through interviews with the user community responsible for developing and maintaining operating procedures for airborne electronic attack systems and with DOD airborne electronic attack research leaders. A more detailed description of our scope and methodology is presented in appendix I.

We conducted this performance audit from February 2011 to March 2012 in accordance with generally accepted government auditing standards. Those standards require that we plan and perform the audit to obtain sufficient, appropriate evidence to provide a reasonable basis for our findings and conclusions based on our audit objectives. We believe that the evidence obtained provides a reasonable basis for our findings and conclusions based on our audit objectives.

BACKGROUND

DOD invests in electronic warfare capabilities as a means to maintain unimpeded access to the electromagnetic spectrum during war and selectively deny adversary use of the spectrum. Traditionally, electronic warfare has been composed of three primary activities:

- **Electronic attack:** Use of electromagnetic, directed energy, or antiradiation weapons to attack with the intent of degrading, neutralizing, or destroying enemy combat capability.
- **Electronic protection:** Passive and active means taken to protect personnel, facilities, and equipment from the effects of friendly or enemy use of the electromagnetic spectrum.
- **Electronic warfare support:** Actions directed by an operational commander to search for, intercept, identify, and locate sources of radiated electromagnetic energy for the purposes of immediate threat recognition, targeting, and planning, and the conduct of future operations.

Airborne electronic attack—a subset of the electronic attack mission—involves use of aircraft to neutralize, destroy, or temporarily degrade (suppress) enemy air defense and communications systems, either through destructive or disruptive means. These capabilities are increasingly important and complex as networked systems, distributed controls, and sophisticated sensors become ubiquitous in military equipment, civilian infrastructure, and commercial networks—developments that complicate DOD's ability to exercise control over the electromagnetic spectrum, when necessary, to support U.S. military objectives.

Airborne electronic attack systems increase survivability of joint forces tasked to enter denied battlespace and engage anti-access threats or high-value targets,[3] whether involved in major combat operations against a potential near-peer adversary or in irregular warfare. They also enable access to the battlespace for follow-on operations. Aircraft executing airborne electronic attack missions employ a variety of mission systems, such as electronic jammers, and weapons, such as antiradiation missiles and air-launched expendable decoys. These aircraft also rely on aircraft self-protection systems and defensive countermeasures for additional protection. All four services within DOD contribute to and rely upon airborne electronic attack capabilities using a variety of different aircraft. Each service is also separately acquiring new airborne electronic attack systems.

Section 1053 of the National Defense Authorization Act for Fiscal Year 2010 requires that for each of fiscal years 2011 through 2015, the Secretary of Defense, in coordination with the Joint Chiefs of Staff and secretaries of the military departments, submit to the congressional defense committees an annual report on DOD's electronic warfare strategy.[4] Each report must contain (1) a description and overview of the department's electronic warfare strategy and organizational structures for oversight; (2) a list and description of all electronic warfare acquisition programs and research and development projects within DOD; and (3) for the unclassified programs and projects, detail on oversight responsibilities, requirements, funding, cost, schedule, technologies, potential redundancies, and associated capability gaps, and for the classified programs and projects, a classified annex addressing these topics, when appropriate.[5] In response to this requirement, DOD submitted its first Electronic Warfare Strategy of the Department of Defense report in October 2010. The department produced its second electronic warfare strategy report in November 2011.

DOD Strategy to Lower Costs Also Reduced Synergy among Systems

DOD's strategy for meeting airborne electronic attack requirements—including both near-peer and irregular warfare needs—centers on acquiring a family of systems, including traditional fixed wing aircraft, low observable aircraft, unmanned aerial systems, and related mission systems and weapons. Department analyses dating back a decade have identified capability gaps and provided a basis for service investments in airborne electronic attack capabilities. However, budget realities and lessons learned from operations in Iraq and Afghanistan have driven changes in strategic direction and program content. Most notably, the department canceled some acquisitions, after which services revised their operating concepts for airborne electronic attack. These decisions saved money, allowing the department to fund other priorities, but reduced the planned level of synergy among airborne electronic attack systems during operations. As acquisition plans for these systems have evolved, operational stresses upon the existing inventory of weapon systems have grown. These stresses have materialized in the form of capability limitations and sustainment challenges for existing systems, prompting the department to invest in improvements to these systems to mitigate shortfalls.

Airborne Electronic Attack Acquisition Strategy Has Evolved

Key DOD analyses completed since 2002 identified capability gaps, provided a basis for service investments in airborne electronic attack systems, and supported an overarching acquisition strategy for achieving these requirements. The department outlined its findings in reports that included an analysis of alternatives, a capabilities-based assessment, and initial capabilities documents. Figure 1 highlights a chronology of these analyses and identifies key airborne electronic attack components of each report.

The 2002 Airborne Electronic Attack Analysis of Alternatives established the primary framework by which the department began investing in new airborne electronic attack capabilities. The analysis focused on those capabilities needed to suppress enemy air defenses from 2010 to 2030. The study identified two primary components required to provide a complete and comprehensive airborne electronic attack solution:

- **Core component:** A recoverable platform or combination of platforms operating in enemy airspace. The core component provides the airborne electronic attack detection and battle management capabilities for reactive jamming.
- **Stand-in component:** An expendable air platform providing critical capabilities against certain advanced threat emitters and employed in threat environments not accessible to the core component.

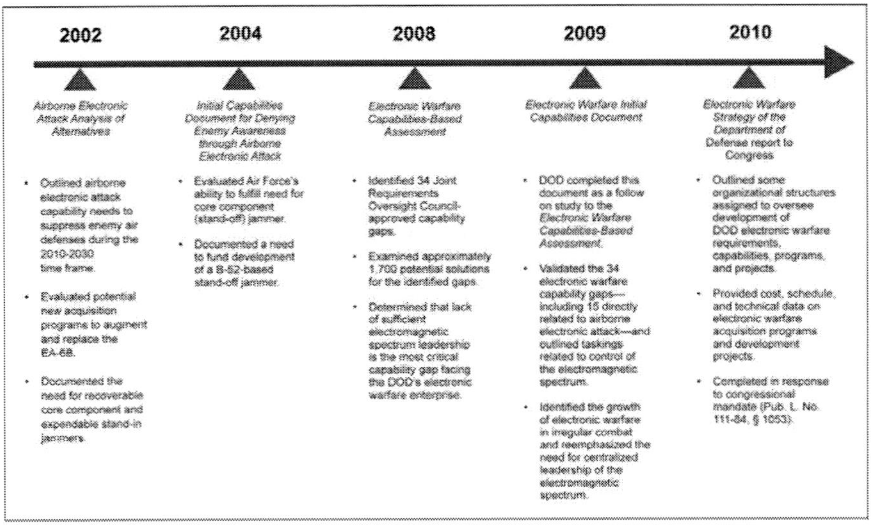

Source: GAO analysis of DOD documentation.

Figure 1. Key Analyses Underpinning Airborne Electronic Attack Acquisition Strategy and Investments.

Subsequent to this analysis, DOD developed a system of systems strategy for meeting airborne electronic attack mission needs. A system of systems is a set or arrangement that results when independent and useful systems are integrated into a larger, connected and interdependent system that delivers unique capabilities during military operations. The system of systems strategy established specific roles and operating responsibilities among the military services in a joint environment and expanded the basic core and stand-in component needs into four major capability areas for airborne electronic attack:

- **Stand-off:** Jamming occurring outside of defended airspace. Planned stand-off systems included the Air Force's EC-130H Compass Call aircraft and development of an electronic attack variant of the Air Force's B-52.
- **Modified escort:** Jamming occurring inside defended airspace, but outside of the range of known surface-to-air missiles. Planned modified escort systems included the Navy's EA-18G Growler and EA-6B Prowler aircraft.

- **Penetrating escort:** Jamming occurring inside the intercept range of known surface-to-air missiles. The department planned to rely on aircraft equipped with active electronically scanned array (AESA) radars, including the F-22A Raptor and F-35 Lightning II aircraft to perform this jamming function.
- **Stand-in:** Jamming occurring inside the "no escape range" of known surface-to-air missiles. The department planned to rely on development of recoverable Joint Unmanned Combat Air Systems (J-UCAS) and the Air Force's Miniature Air Launched Decoy—Jammer (MALD-J) to provide this function.

As time progressed, budget issues and lessons learned from operations in Iraq and Afghanistan drove changes to the strategy and program content. Most notably, the department canceled development of two major components of the system of systems—the B-52 Standoff Jammer and J-UCAS—in 2005 and 2006, respectively, citing higher-priority needs and budget constraints. The B-52-based jamming concept was later rejuvenated through the Air Force's Core Component Jammer initiative, but that program was similarly canceled in 2009. Following these developments, the department revised operating concepts and joint service responsibilities, moving away from its system of systems plans in favor of a family of systems strategy for airborne electronic attack.

A family of systems is fundamentally different from a system of systems. Under a family of systems construct, independent systems—using different approaches—together provide capability effects to support military operations. Unlike the synergy found in a system of systems, a family of systems does not acquire qualitatively new properties or necessarily create capability beyond the additive sum of the individual capabilities of its members. The member systems may not even be connected into a whole. In the case of airborne electronic attack, DOD officials stated that a system of systems would have employed a dynamic, networked capability to share data in real-time among platforms—a concept known as electronic warfare battle management. Under the family of systems strategy, officials stated that this process is less automated and the parts are less connected. Therefore, in making this strategy change, the department traded some unique, synergistic capabilities that the system of system's interdependent components might have provided in favor of near-term budget savings and other priorities.

Figure 2 outlines the department's current family of systems strategy for countering near-peer adversaries. This family of systems includes traditional fixed wing aircraft, low observable aircraft, and related mission systems and weapons.

DOD's 2009 electronic warfare capabilities analysis identified the growth of irregular warfare in urban areas as presenting challenges to military operations. The analysis noted that irregular adversaries can exploit civilian and commercial communications infrastructure to minimize detection and subsequent attack. According to the department, precise electronic attack planning and execution are required to ensure that these threats are defeated while avoiding interruption to U.S. communications capabilities.

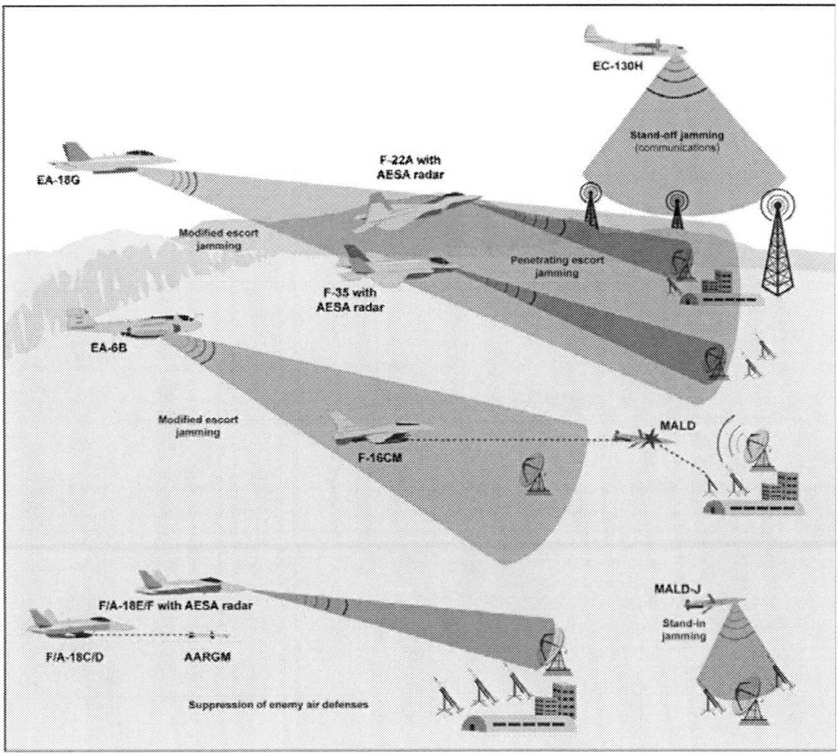

Source: GAO (presentation); Northrop Grumman (EA-6B); Boeing (EA-18G); Raytheon (MALD and MALD-J); Lockheed Martin (F-35); Department of Defense (EC-130H, F-22A, AARGM, F-16CM, F/A-18C/D, and F/A-18E/F); Art Explosion (all other images).

Figure 2. Airborne Electronic Attack Family of Systems Strategy for Countering Near-Peer Adversaries.

The department has used existing airborne electronic attack systems, such as the EA-6B and EC-130H, to meet its near-term irregular warfare needs in Iraq and Afghanistan. However, officials report that these platforms are optimized for countering high-end, near-peer threats, and their use against irregular warfare threats is inefficient and costly. Consequently, the department has begun investing in new, less expensive airborne electronic attack systems tailored to counter irregular warfare threats. These systems are fielded from both traditional fixed-wing aircraft and from unmanned aerial vehicles. Figure 3 illustrates operations involving these systems.

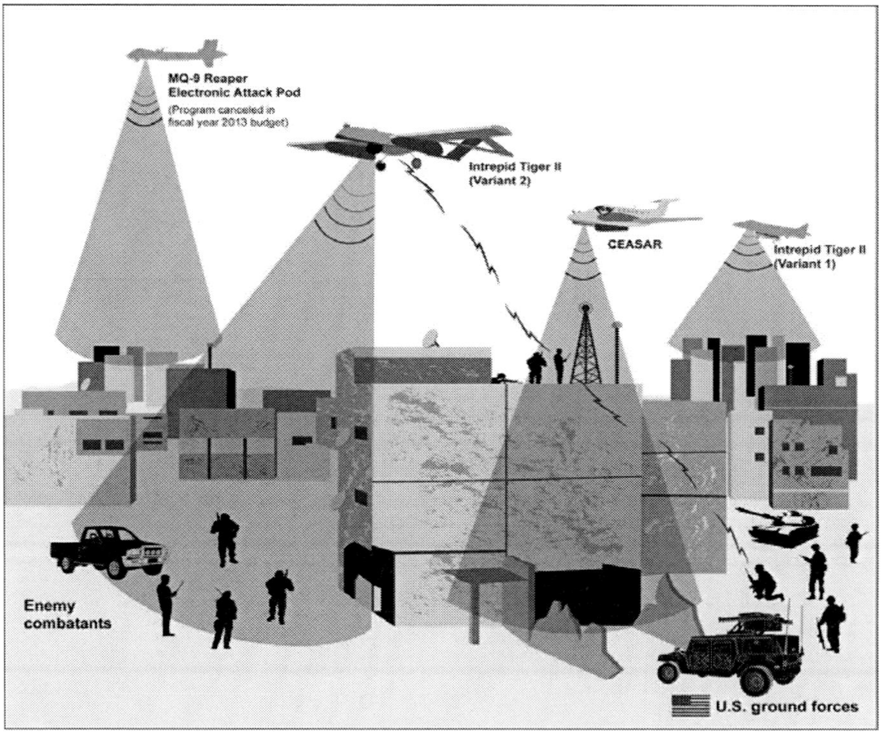

Source: GAO (presentation); Department of Defense (MQ-9 Reaper, CORPORAL, CEASAR, and Intrepid Tiger II); Art Explosion (all other images).

Figure 3. Airborne Electronic Attack Systems Tailored to Counter Irregular Warfare Threats.

Existing Airborne Electronic Attack Systems Face Capability Limitations and Sustainment Challenges

As DOD's acquisition plans for airborne electronic attack systems have evolved, operational stresses upon the current inventory of systems have grown. These systems date back to the 1970s and 1980s and were originally designed to counter Cold War era threats. Many of the department's existing airborne electronic attack systems face capability limitations, requiring the department to pursue modernization efforts to increase the effectiveness of the systems or to identify and develop replacement systems. Further, existing systems face sustainment challenges from age, parts obsolescence, and increased operational stresses from lengthy and sustained operations in Iraq and Afghanistan. According to Air Force and Navy officials, these challenges have reduced the availabilities of some systems to warfighters. Table 1 identifies the department's existing airborne electronic attack systems and related characteristics, including future replacement systems identified to date.

DOD is taking actions to address capability limitations and sustainment challenges across several key systems, such as the following:

- **EA-6B Prowler:** Since its introduction in the 1970s, the Navy and Marine Corps have made significant upgrades to the EA-6B Prowler. The latest of these upgrades—the Improved Capability electronic suite modification (ICAP III) provides the Prowler with greater jamming capability and is designed to improve the aircraft's overall capability as both a radar-jamming and HARM platform. By the end of fiscal year 2012, 32 EA-6Bs will be upgraded to the ICAP III configuration. Navy officials told us that persistent operations in Iraq and Afghanistan, however, have degraded the condition of EA-6B aircraft. In addition, we have previously reported that parts obsolescence presents the biggest challenge to the EA-6B's ability to fulfill its mission role.[6] We noted that although the Navy has made several structural upgrades to the EA-6B fleet, it is actively tracking a number of key components, including cockpit floors, side walls, fin pods, bulkheads, actuators, engine components, landing gear, and avionics software—all of which are at increasing risk for costly replacement the longer the aircraft remains in service.
- **HARM:** According to Navy officials, even though HARM has undergone various block upgrades to provide increased capabilities since fleet introduction in 1983, advancements in enemy radar

technology have rendered the weapon somewhat ineffective for typical Navy targets. As a result, the Navy is fielding a major technological upgrade to HARM through its AARGM acquisition program. AARGM provides a new multimode guidance section and modified control section mated with existing HARM propulsion and warhead sections. The Air Force, similarly, is pursuing modifications to HARM control sections on missiles in its inventory—a process that will provide a global positioning system receiver to those units. Air Force officials stated that they have long sought this receiver component addition because of vulnerabilities in the HARM targeting method. This effort is being pursued in conjunction with other modernization efforts for Air Force F-16CM aircraft.

Table 1. Characteristics of Airborne Electronic Attack Systems in Sustainment

System	Mission description	Estimated end of service life	Replacement system
EA-6B Prowler	Modified escort jamming	2020	EA-18G (Navy) F-35B Lightning II (Marine Corps)[a]
AN/ALQ-99 Tactical Jamming System	Modified escort jamming	Mid-band: 2024 Low-band: 2026 High-band: 2028	Next Generation Jammer
F-16CM	Suppression of enemy air defenses	2024	F-35A Lightning II
AN/ALQ-131 and AN/ALQ-184 Pod Systems	Aircraft self-protection (F-16 and A-10)	2025	Electronic Attack Pod Upgrade Program
AN/ALQ-135 Internal Countermeasures Systems	Aircraft self-protection (F-15)	2035	Eagle Passive/Active Warning Survivability System
AGM-88 High Speed Anti-Radiation Missile (HARM)	Suppression of enemy air defenses	2035[b]	Advanced Anti-Radiation Guided Missile (AARGM)
EC-130H Compass Call (Baselines 0 and 1)	Stand-off jamming (communications)	2053[c]	N/A
ADM-141 Tactical Air Launched Decoy (TALD)/ Improved Tactical Air Launched Decoy (ITALD)	Suppression of enemy air defenses	Unknown[d]	TBD[e]

System	Mission description	Estimated end of service life	Replacement system
F-22A Raptor	Penetrating escort	TBD	N/A
Integrated Defensive Electronic Counter-measures (IDECM) Blocks 1 and 2	Aircraft self-protection (F/A-18 E/F)	TBD	IDECM Blocks 3 and 4

Legend: N/A = not applicable; TBD = to be determined.
Source: GAO analysis of DOD data.

[a] In addition to the fixed wing, airborne electronic attack capability that F-35B Lightning II is anticipated to provide, the Marine Corps plans to rely on its Marine Air Ground Task Force Electronic Warfare concept to replace the warfighting capability and capacity currently provided by the EA-6B. This concept seeks a more holistic approach toward electronic warfare by combining both air and ground capabilities. To date, DOD officials state that the Marine Corps has completed a draft initial capabilities document, a concept of operations, and various electronic warfare gap analyses in support of its concept.

[b] This date refers to the expected service life of the Air Force's inventory of HARM only. Air Force officials told us that retirement of the Air Force's inventory of HARM is aligned with the expected retirement of Block 50/52 F-16 aircraft.

[c] As of January 2012, the EC-130H program schedule showed that center wing box replacement for the 14th Compass Call aircraft should be complete by 2018. A program office official told us that center wing box replacement extends the operational service life of the aircraft an additional 35 years, suggesting an end of service life in 2053, assuming no additional improvements to the fleet.

[d] According to a Navy official, neither TALD nor ITALD has an estimated end of service life. The Navy plans to continue providing minimal sustainment funds for these systems, as resource availability permits.

[e] The Navy has begun evaluating TALD/ITALD replacement options.

- **TALD and ITALD:** Navy officials stated that advancements in enemy integrated air defense systems have decreased the effectiveness of both TALD and ITALD units. According to program officials, newer radars can discern from the TALD/ITALD flight profile that the system is a decoy and not a valid target. The Navy has begun evaluating TALD/ITALD replacement options under its Airborne Electronic Attack Expendable program initiative.
- **EC-130H Compass Call (Baselines 0 and 1):** Although the Air Force initially fielded the EC-130H Compass Call as a communications jammer supporting suppression of enemy air defenses, the system has evolved to include irregular warfare missions and radar jamming.

  Air Force officials told us that the Compass Call is the most utilized aircraft within the C-130 family and has been continuously deployed since 2003 supporting operations in Iraq and Afghanistan, accelerating the need for the Air Force to replace the center wing box on each of the 14 aircraft in the Compass Call fleet. Further, Air Force officials told us that they are increasing the size of the fleet by one aircraft to alleviate stress on current aircraft and to increase the availability of airborne electronic attack capability to the Air Force. According to a fleet viability assessment completed in 2010, the current size of the fleet is insufficient to meet combatant commander taskings for Compass Call.

- **AN/ALQ-99 Tactical Jamming System:** The Navy's Low Band Transmitter upgrade to the AN/ALQ-99 system is intended to replace three aging legacy transmitters that suffer from obsolescence and reliability problems. According to Navy officials, persistent use of these transmitters in support of operations in Iraq and Afghanistan has exacerbated system shortfalls. Navy officials told us that they are also identifying options for improving reliability and resolving obsolescence issues with the mid and high bands of the AN/ALQ-99 system.
  However, Navy officials project that even with these improvements, system capabilities will be insufficient to counter anticipated evolutions in threat radars and missiles beginning in 2018. This shortfall is expected to be addressed by the new Next Generation Jammer.
- **AN/ALQ-131 and AN/ALQ-184 Pod Systems:** The Air Force has identified obsolescence issues and capability shortfalls affecting these systems, which provide tactical aircraft self-protection. The Air Force is pursuing a replacement/upgrades program designed to move the Air Force to a single, self-protection pod system for its F-16 and A-10 aircraft.

ACQUISITIONS MAY NOT PRODUCE SUFFICIENT RESULTS

DOD is investing in new airborne electronic attack systems to address its growing mission demands and to counter anticipated future threats. However, progress acquiring these new capabilities has been impeded by developmental and production challenges that have slowed fielding of several planned systems. Some programs, including the Navy's EA-18G Growler and the Air

Force's EC-130H Compass Call modernization, are in stable production and have completed significant amounts of testing. On the other hand, the Navy's AARGM, the Air Force's Miniature Air Launched Decoy (MALD), and other programs have required additional time and money to resolve technical challenges. In addition, certain airborne electronic attack systems in development may offer capabilities that overlap with one another—a situation brought on in part by the department's fragmented urgent operational needs processes. As military operations in Iraq and Afghanistan decrease, opportunities exist to consolidate current acquisition programs across services; however, this consolidation may be hampered by leadership deficiencies affecting the department's electronic warfare enterprise. Furthermore, current and planned acquisition programs, even if executed according to plan, will not fully address the materiel-related capability gaps identified by the department—including some that date back 10 years.

Investments in New Airborne Electronic Attack Systems Have Yielded Mixed Results to Date

DOD investments to develop and procure new and updated airborne electronic attack systems are projected to total more than $17.6 billion from fiscal years 2007 through 2016.[7] These systems represent the department's planned mix of assets for (1) countering near-peer, integrated air defense and communications systems and (2) providing communications and radio frequency jamming against irregular warfare threats. Table 2 outlines the department's recent and planned investments toward developing and acquiring several of these systems. As table 2 shows, several airborne electronic attack systems are in an advanced stage of funding. However, under current estimates, over $6.0 billion in funding is still required to fully deliver these new systems to the warfighter. Further, the department has not yet identified the full amount of funding required for certain key systems, such as the Next Generation Jammer, which could require billions of additional dollars to field. Correspondent to their different funding profiles, the department's new systems are also in various stages of development, with some progressing more efficiently than others. Table 3 identifies the mission role(s), developmental status, and fielding plans for these systems. In addition, appendix II provides additional details on the status of several of these programs.

Table 2. Recent and Planned DOD Investments toward Acquiring Airborne Electronic Attack Systems

Then-year dollars in millions

System	Current research, development, testing, and evaluation (RDT&E) cost estimate	Current procurement cost estimate	Total RDT&E and procurement funding (through fiscal year 2012)	Remaining RDT&E and procurement funding requirements
Next Generation Jammer	$2,141.5	N/A	$402.6	$1,738.9
EA-18G Growler	1,839.4	$9,341.6	10,032.5	1,148.5
AARGM	631.0	1,277.7	877.5	1,031.2
MALD and MALD-J[a]	505.1	1,339.6	953.8	890.9
IDECM Block 4	254.1	609.1	271.6	591.6
EC-130H Compass Call Modernization	129.0	957.2	709.9	376.3[b]
MQ-9 Reaper Electronic Attack Pod	133.4	100.3	53.1	0.0[c]
Intrepid Tiger II	26.5	50.3	43.4	33.4
Communications Electronic Attack with Surveillance and Reconnaissance (CEASAR) pod	0.8	13.0	13.8[d]	0.0
Total	**$5,660.8**	**$13,688.8**	**$13,358.2**	**$5,810.8**

Source: GAO analysis of DOD budget submissions and program baselines.

Notes: Remaining funding requirements for EC-130H Compass Call Modernization represent funding through the end of a 5-year budget forecast. In addition, F-22A Raptor and F-35 Lightning II (Joint Strike Fighter) funding is excluded from this analysis because those aircraft will provide capabilities that support several missions, including airborne electronic attack.

[a] MALD and MALD-J figures do not include costs or appropriations related to the previously planned MALD-J Increment II. In its fiscal year 2013 budget submission, the Air Force canceled the Increment II program. Prior to this cancellation, the Air Force planned to invest $272.3 million in RDT&E funding through fiscal year 2016 to begin developing this new capability.

[b] Total does not include funding needed to support Air Force plans to modernize three additional EC-130H aircraft in fiscal years 2017 through 2018 because the Air Force has not yet identified these funding requirements. However, according to Air Force officials, they expect the 2017 and 2018 modernization budgets to each remain constant with planned fiscal year 2016 funding of $85.5 million.

[c] In its fiscal year 2013 budget submission, the Air Force canceled the MQ-9 Reaper Electronic Attack Pod program. Prior to this cancellation, remaining funding requirements for the program were expected to total $180.6 million.

[d] Total excludes approximately $16.0 million in program funding from the Operations and Maintenance, Army account, which the Army has used, in part, to lease C-12 aircraft to host the CEASAR pod. In fiscal year 2013, the Army plans to request an additional $10.3 million in Operations and Maintenance funding to support these activities.

Table 3. DOD's Progress Developing and Fielding New Airborne Electronic Attack Systems

System	Mission description	Development status	Actual/estimated fielding date
EA-18G Growler	Modified escort jamming	The Navy has fielded EA-18G aircraft with limited cost and schedule growth to date. Through fiscal year 2011, the Navy placed 90 of the planned 114 Growler aircraft under production contract. Operational testing identified suitability concerns, which the Navy has addressed through software changes. In July 2011, the system completed its first combat deployment supporting operations in Iraq and Libya. The Navy continues to develop new software blocks to add aircraft capability.	2009
CEASAR pod	Irregular warfare jamming	In September 2011, the Army initiated an operational assessment of the CEASAR system by deploying three pods; two contractor-owned, government-operated C-12 aircraft; and associated operators and support personnel to Operation Enduring Freedom. Following this authorized 1-year assessment, the Army will make a determination on CEASAR's readiness to transition into a formal acquisition program.	2011
Large Aircraft Infrared Counter-measures (LAIRCM)	Aircraft self-protection	Current acquisition plans add a next generation missile warning system to LAIRCM to provide improved detection against infrared threat missiles. Recently, the next generation missile warning system completed initial operational test and evaluation, and a full rate production decision is planned for 2012.	2011
IDECM Blocks 3 and 4	Aircraft self-protection	IDECM Block 3 entered full rate production in 2011. IDECM Block 4 integrates significant hardware design changes to the ALQ-214 onboard jammer component. These changes will enable the system to operate on F/A-18C/D aircraft, while maintaining the system's functionality on F/A-18E/F aircraft. Planned concurrency in the Block 4 testing and production schedules increases risk of retrofits to delivered systems.	2011 (Block 3) 2014 (Block 4)

Table 3. (Continued)

System	Mission description	Development status	Actual/estimated fielding date
AARGM	Suppression and destruction of enemy air defenses	Hardware and software failures during operational testing in 2010—and subsequent deferral of remaining testing—drove a 9-month fielding delay to the system. Manufacturing quality and reliability concerns prompted the Navy to institute a "fly before you buy" program to screen poor weapons prior to government acceptance. AARGM recently resumed operational testing, but the Navy assesses system suitability as high risk.	2012
Intrepid Tiger II	Irregular warfare jamming	The Intrepid Tiger II program is developing 2 pod variants for AV-8B (variant 1) and RQ-7B (variant 2) aircraft. Variant 1 entered operational testing in 2011 ahead of planned deployment of initial pods. Design change costs, including a radio system change, were absorbed by reducing pod quantities (14 to 8). Variant 2 testing under the Collaborative On-line Reconnaissance Provider/Operationally Responsive Attack Link demonstration program concluded in April 2011. Full performance of variant 2 pods remains unproven due to platform (RQ-7B) unavailability and integration issues, including susceptibility to electromagnetic interference.	2012 (Variant 1) TBD (Variant 2)
MALD	Suppression of enemy air defenses	MALD operational testing was suspended following anomalies and subsequent crashes of test vehicles in June 2010 and February 2011. System design changes facilitated a return to testing, but an August 2011 test shot also failed. Testing concluded in September 2011.	2012
MALD-J	Stand-in jamming	MALD-J employs the same flight vehicle as MALD, with slight differences to account for inclusion of a jammer. The Air Force approved MALD-J entry into low rate initial production in September 2011, with planned production start in May 2012. Operational testing has been reduced from 15 to 7 months in an attempt to mitigate program delays resulting from MALD design deficiencies. DOD states this reflects an increase in test range priority and a decrease in data turnaround time.	2012
MALD-J Increment II	Stand-in jamming	Funding shortfalls curtailed Air Force plans to award a technology development contract in fall 2011 for MALD-J sensitivity and jamming power improvements—key capability gains intrinsic to Increment II. The program was later canceled In the fiscal year 2013 budget submission.	N/A

System	Mission description	Development status	Actual/estimated fielding date
EC-130H Compass Call (Baselines 2 and 3)	Stand-off jamming (communications)	Baseline 2 modernization—currently scheduled for 8 of the Air Force's planned 15 EC-130H aircraft—adds new capabilities including improved special purpose emitter array transmitters and addresses aircraft obsolescence issues. Modification work on the first of these 8 aircraft began in fiscal year 2011, with 3 more aircraft following in fiscal year 2012. The Air Force is currently studying configuration options for Baseline 3, which officials expect to install on 7 EC-130H aircraft.	2014 (Baseline 2) 2017 (Baseline 3)
F-35 Lightning II (Joint Strike Fighter)	Penetrating escort	F-35 entered low rate initial production in 2007 and has since experienced significant cost growth and schedule delays. Development challenges caused the program to be restructured in 2010, triggering a Nunn-McCurdy cost breach.[a] GAO has repeatedly expressed concerns about the F-35's technology maturity and design stability. The program revised its testing plan and is making progress against a new schedule.	TBD[b]
MQ-9 Reaper Electronic Attack Pod	Irregular warfare jamming	Prior to cancelling the program in its fiscal year 2013 budget submission, the Air Force planned to integrate electronic attack pods on Block 5 MQ-9 aircraft—the first units expected to have sufficient power to operate the pods. Program officials stated that electromagnetic interference caused by the pods jamming the MQ-9 command and control systems posed a key technical challenge. The program entered technology maturation phase in 2010 and planned to award an engineering and manufacturing development contract in 2013.	N/A
Next Generation Jammer	Modified escort jamming	The Navy plans to award technology development contracts for the system in the third quarter of fiscal year 2013, with award of an engineering and manufacturing development contract to follow in 2015. In November 2010—based on findings from the system's analysis of alternatives—Navy leadership directed the program to pursue a block approach to developing capability, whereby mid-, low-, and high-band jammers would be progressively fielded on EA-18G aircraft and, through a later increment, F-35 aircraft.	2020 (Mid-band) 2022 (Low-band) 2024 (High-band)[c]

Legend: N/A = not applicable; TBD = to be determined.

Source: GAO analysis of DOD data.

[a] A Nunn-McCurdy cost breach occurs when a program's unit cost exceeds certain statutory thresholds.

[b] F-35 does not currently have an approved fielding date. Prior to the program's Nunn-McCurdy breach, the Marine Corps planned to declare initial operational capability for the aircraft in 2012. However, since the breach, DOD has not yet approved a new acquisition program baseline, and the services continue to evaluate potential fielding dates for the F-35.

[c] Dates provided reflect system fielding plans with EA-18G; fielding dates with F-35 are undetermined.

Some Programs Are Progressing Well

Some airborne electronic attack acquisition programs have reached stable production with limited cost growth or schedule delays. Two primary examples include the following:

- **EA-18G Growler:** Acquisition of the EA-18G Growler—a modified escort jamming platform designed to carry AN/ALQ-99 and future Next Generation Jammer pods—achieved initial capability in September 2009, consistent with its 2007 baseline schedule. Additionally, program costs per aircraft increased less than one-half of 1 percent from 2003 to 2010—an outcome partially attributable to quantity increases from 90 to 114.
- **EC-130H Compass Call (Baselines 2 and 3):** Modernization of the EC-130H Compass Call is on schedule for fielding a new increment of capability, Baseline 2, in 2014 within available funding limitations. Baseline 2 introduces several new capabilities, including reactive radar response and the Joint Tactical Radio System terminal that has been delayed because of testing challenges. However, Compass Call program officials do not expect the radio system delay to affect the program's fielding plans for Baseline 2 aircraft. According to the Air Force, cost considerations are a primary criterion in developing EC-130H capability requirements. The program office does not entertain potential aircraft improvements unless those improvements are accompanied by full funding. The Air Force is initiating technology development activities for a subsequent phase of the modernization program, Baseline 3, and plans to begin production of these aircraft in 2014, with initial fielding scheduled for 2017.

Our previous work has shown that good acquisition outcomes are achieved through the knowledge-based approach to product development that demonstrates high levels of knowledge before significant commitments are made.[8] In essence, knowledge supplants risk over time.

This model relies on increasing knowledge when developing new products, separating technology development from product development, and following an evolutionary or incremental approach to product development.

In this approach, developers make investment decisions on the basis of specific, measurable levels of knowledge at critical junctures before investing more money and before advancing to the next phase of acquisition. The good outcomes on the EA-18G and EC-130H programs can be attributed, in part, to acquisition strategies embodying elements of best practices.

Some Programs Are Underperforming

Other airborne electronic attack acquisition programs have not progressed as efficiently, however. These systems have proceeded through product development with lower-than-desired levels of knowledge and subsequently faced technical, design, and production challenges, contributing to significant cost growth, fielding delays or both.

Most notably, these systems entered—or are on track to enter—production before completing key development activities, including achievement of stable designs.

We previously reported that concurrency in development and production activities limits the ability of an acquisition program to ensure that the system will work as intended and that it can be manufactured efficiently to meet cost, schedule, and quality targets.[9]

- **MALD/MALD-J:** MALD was authorized for low rate initial production in June 2008 with an initial plan for 300 low rate initial production units in two lots, beginning in March 2009. However, testing failures in 2010 and 2011—coupled with a desire to avoid a potentially costly break in production—prompted the Air Force to extend MALD low rate initial production by two additional lots and increase total quantities under contract to 836. In September 2011, citing "successful completion of MALD-J engineering and manufacturing development activities," the Air Force exercised a priced option to upgrade 240 of its planned MALD units to the MALD-J configuration, subsequently decreasing MALD quantities to 596. Because all future production lots are now planned as jammer-configured decoys (MALD-J), the 596 total represents the full MALD procurement—without the program having ever met the criteria necessary to proceed into full rate production.[10] Since the MALD and MALD-J designs are identical—except for the addition of a jammer module to MALD-J—the absence of a proven manufacturing process for MALD introduces schedule risk to production of MALD-J.[11] This risk is accentuated by continuing deficiencies affecting the MALD and MALD-J designs, which have required the Air Force to schedule additional developmental flight tests for each system in February 2012 to test corrective fixes. To the extent that this retesting phase shows a need for additional design changes, the Air Force may be forced to revisit its planned May 2012 production start for MALD-J.

- **AARGM:** The Navy authorized low rate initial production of AARGM units in September 2008 with initial deliveries scheduled to begin in January 2010. A total procurement objective of 1,919 units was set and an initial operational capability scheduled for March 2011. However, as a result of intermittent hardware and software failures in testing, the program was decertified for initial operational test and evaluation in September 2010, and low rate initial production deliveries were delayed until June 2011. The missile has subsequently reentered testing, but significant concerns about the system's reliability remain. Further, Navy officials stated that the current program schedule is oriented toward success with virtually no margin to accommodate technical deficiencies that may be discovered during operational testing. In the event operational testing reveals new or lingering major deficiencies, program officials report the planned April 2012 fielding date will be at risk, and the Navy may be forced to revisit its commitment to the program.
- **IDECM:** From December 2000 to June 2010, the Navy authorized six different low rate initial production lots of IDECM Blocks 2 and 3, providing system improvements to the jammer and decoy components. Block 2 production units delivered ahead of schedule, but early Block 3 units encountered operational testing failures; later resolved, these failures drove production delays to remaining units. In Block 4, the Navy is introducing significant hardware design changes to the ALQ-214 jammer component. Ground and flight testing to prove out these design changes is scheduled concurrent with transition to production in April 2012, increasing risk that initial Block 4 units will require design changes and retrofits.[12] Officials stated that this concurrency is necessary in order to maintain an efficient production line transition from Block 3 to Block 4 and to meet the desired June 2014 fielding date. They further noted that transition to Block 4 production will initially be for 19 systems, with production rates increasing to as many as 40 per year following completion of testing.

Planned Systems May Offer Capabilities That Overlap, Presenting Opportunities to Consolidate Acquisition Efforts

Certain airborne electronic attack systems in development may offer capabilities that unnecessarily overlap with one another. This condition appears most prevalent with irregular warfare systems that the services are acquiring under DOD's fragmented urgent operational needs processes. For example, the Marine Corps, Army, and Air Force have all separately invested to acquire unique systems intended to jam enemy communications in support of ground forces. Further, Navy and Air Force plans to separately invest in new expendable decoy jammers—systems intended to counter near-peer adversaries—also appear to overlap. Declining military operations in Iraq and Afghanistan—coupled with recent changes in the Air Force's MALD-J program—afford opportunities to consolidate current service-specific acquisition activities. The department's ability to capitalize on these opportunities, however, may be undermined by a lack of designated, joint leadership charged with overseeing electronic warfare acquisition activities.

Potential Overlap among Irregular Warfare Systems Driven by Service-Specific Solutions to Urgent Warfighting Needs

DOD is investing millions of dollars to develop and procure airborne electronic attack systems uniquely suited for irregular warfare operations. Services are acquiring these systems under both rapid acquisition authorities as well as through the traditional acquisition process. These systems overlap—at least to some extent—in terms of planned mission tasks and technical challenges to date. Yet, they have been developed as individual programs by the different services. Table 4 highlights overlap among three of these systems.

According to DOD officials, airborne electronic attack limitations in recent operations, urgent needs of combatant commanders, and the desire to provide ground units with their own locally controlled assets have all contributed to service decisions to individually develop their own systems to address irregular warfare threats. For example, one Marine Corps official told us that his service is focused on increasing its airborne electronic attack capacity to meet Marine Air-Ground Task Force requirements in combat. Marine Corps systems typically equipped to perform these tasks—especially the EA-6B Prowler aircraft—have reached their capacity limits responding to combatant commander taskings. Similarly, Air Force officials stated that ground warfighter requests for airborne electronic attack capabilities sometimes go unfulfilled or are delayed because of the overall constrained capacity during current operations.

Further, Army and Marine Corps officials see operational benefits to providing ground unit commanders with smaller airborne electronic attack assets—permanently integrated within the unit—to free up Air Force and Navy assets for larger-scale missions.

In addition, the capabilities offered by current jamming pods, such as the AN/ALQ-99, are often overkill for the irregular warfare mission needs—such as counter-improvised explosive device activities—facing ground unit commanders.

Requirements for several of these irregular warfare systems were derived from DOD urgent needs processes—activities aimed at rapidly developing, equipping, and fielding solutions and critical capabilities to the warfighter in a way that is more responsive to urgent requests than the department's traditional acquisition procedures.

As we previously reported, the department's urgent needs processes often lead to multiple entities responding to requests for similar capabilities, resulting in potential duplication of efforts.[13] Even under these circumstances, the services have shown it is possible to take steps to share technical information among the different programs and services. For instance, the Army's CEASAR pod is derived from the AN/ALQ-227 communications jammer used on the Navy's EA-18G—an attribute that Army officials state reduced design risk in the program and provided opportunities for decreased sustainment costs and reuse of jamming techniques between the two services. Similarly, Air Force efforts to develop electronic attack pods flown on MQ-9 Reaper unmanned aerial vehicles (prior to that program's cancellation) leveraged previous technology investments for the canceled B-52-based stand-off jammer.

As military operations in Iraq and Afghanistan wind down—and the services evaluate whether to transition their current urgent needs programs over to the formal weapon system acquisition process—opportunities may exist to consolidate program activities, such as the Intrepid Tiger II and CEASAR systems that are still demonstration programs whose transitions to formal acquisition programs have not yet been determined.

Table 4. Potential Overlap among Communications Jamming Systems Supporting Ground Forces

System name	Intrepid Tiger II	CEASAR Pod	MQ-9 Reaper Electronic Attack Pod
Service sponsor	Marine Corps	Army	Air Force
Host platform	Variant 1: AV-8B fixed wing aircraft[a] Variant 2: RQ-7B unmanned aerial vehicle	C-12 fixed wing aircraft	MQ-9 Reaper unmanned aerial vehicle
Mission description	Communications jamming and surveillance capability in support of ground forces	Denial and disruption of enemy communications systems and improvised explosive devices in support of unit-level ground commanders	Communications and improvised explosive device jamming in support of combatant commander mission needs
Technical status	Program recently completed compatibility testing for variant 1 to identify potential electromagnetic interference issues and reduce system fratricide. Interoperability testing will not be completed until after the system has achieved early operational capability (fielding). Initial testing of variant 2 revealed electromagnetic interference with the RQ-7B's safety of flight systems as well as aircrew system feedback and usability issues with the electronic attack payload system interface.	Electromagnetic Interference issues—resulting from continuous low frequency jamming—were identified in testing, subsequently causing impairment to aircraft navigation and communications systems. According to Army officials, these challenges have been overcome with solutions proven during the system's recent 2011 operational assessment.	The Air Force canceled this program in its fiscal year 2013 budget submission. Prior to cancellation, program officials anticipated potential MQ-9 electromagnetic interference issues caused by the jamming pod that could interfere with the aircraft's communications link to ground station controllers.
Estimated acquisition cost	$76.8 million	$13.8 million[b]	$233.7 million[c]

Source: GAO analysis of DOD data.

[a] Integration and fielding on AV-8B aircraft represent minimum (threshold) requirements for the Intrepid Tiger II (Variant 1) pod. Beginning in fiscal year 2012, the Marine Corps plans to transition Intrepid Tiger II (Variant 1) to other fixed and rotary wing aircraft, including the F/A-18C/D.

[b] Total excludes $26.3 million in funding from the Operations and Maintenance, Army budget account through fiscal year 2013. The Army uses these funds to (1) lease two C-12 aircraft to fly the CEASAR pod and (2) fund aircraft and pod sustainment costs.

[c] Reflects estimated acquisition cost prior to program cancellation.

Navy and Air Force Have Not Agreed on a Common Decoy Solution

The potential for unnecessary overlap in efforts within the airborne electronic attack area is not limited to irregular warfare systems. With respect to near-peer systems, both the Air Force and Navy are separately pursuing advanced jamming decoys—the Air Force through its MALD-J program, and the Navy through its planned Airborne Electronic Attack Expendable initiative.

The two services have held discussions with one another about combining efforts toward a joint solution, including a meeting between Navy and Air Force requirements offices and acquisition officials in December 2010, but they have not yet reached resolution on a common path forward. According to Navy officials, relatively minor design and software modifications to what was a planned second increment to the Air Force's MALD-J system could produce a system that satisfies both services' mission requirements. However, Air Force officials stated that accommodating the Navy's mission requirements within the system would increase program costs and delay planned fielding of the Increment II system, essentially rendering the planned program unexecutable. Subsequently, Air Force officials stated that unless Increment II, in its planned configuration, sufficiently met Navy requirements, they did not expect the Navy to have any formal role in the program. In July 2011, however, the Air Force suspended MALD-J Increment II activities because of a lack of future funding availability. In February 2012, the Air Force's fiscal year 2013 budget submission officially canceled the program.[14] This cancellation affords an opportunity for continued dialogue between the two services on the potential benefits and drawbacks to pursuing a common acquisition solution.

Leadership Deficiencies Undermine the Department's Ability to Reduce Overlap

In 2009, DOD completed a capabilities analysis that cited electromagnetic spectrum leadership as the highest priority among 34 capability gaps identified. The study concluded, in part, that leadership deficiencies, or its absence, significantly impede the department from both identifying departmentwide needs and solutions and eliminating potentially unnecessary overlap among the services' airborne electronic attack acquisitions. Specifically, the department lacks a designated, joint entity to both coordinate internal activities and represent electronic warfare activities and interests to outside organizations. Acknowledging this leadership gap, and its relation to acquisition activities, the department has initiated efforts to organize the Joint

Electromagnetic Spectrum Coordination Center under the leadership of U.S. Strategic Command. In addition, officials representing the Office of the Assistant Secretary of Defense for Research and Engineering stated that they are considering actions they might take to improve leadership and oversight of electronic warfare acquisition activities across the services. In a separate report, we intend to evaluate planned and existing electronic warfare governance structures within DOD.

Planned Systems Will Not Fully Address Capability Gaps

Notwithstanding the considerable investment over the years in new and enhanced airborne electronic attack systems and subsystems, capability gaps, some identified a decade ago, are expected to persist, or even increase, through 2030 as adversary capabilities continue to advance. In a series of studies since 2002, DOD identified existing current and anticipated gaps in required capabilities. Some have persisted for years—for example, deficiencies in certain jamming capabilities to provide cover for penetrating combat aircraft. The analyses found that, in many cases, new materiel solutions were required to close these gaps. Table 5 outlines primary findings from three major analyses.

Table 5. Primary Airborne Electronic Attack Capability Needs Identified since 2002

Capabilities analysis	Analysis sponsor	Needs identified
Airborne Electronic Attack Analysis of Alternatives (2002)	Office of the Secretary of Defense	Stand-in and core component jamming capability needs identified. The analysis outlined 27 potential platform combinations to address these needs.
Initial Capabilities Document for Denying Enemy Awareness through Airborne Electronic Attack (2004)	Air Force	Identified Air Force needs for materiel solutions to provide stand-off and modified escort jamming, in light of the then-pending retirement of Navy EA-6B Prowler aircraft. The document also identified penetrating escort and stand-in jamming capability needs unique to the Air Force, while identifying potential materiel solutions.
Electronic Warfare Initial Capabilities Document (2009)	U.S. Strategic Command	Identified 34 electronic warfare enterprise-wide capability gaps. Fifteen of these gaps relate directly to the airborne electronic attack mission area. The study concluded that of these 15 gaps, 7 require new materiel solutions. Top priority is fixing leadership shortfalls.

Source: GAO analysis of DOD data.

The 2002 analysis identified needs for stand-in and core component jamming capabilities and suggested numerous ways to meet these. The 2004 study revalidated these gaps and outlined 10 potential materiel solutions to fill those gaps. It also acknowledged the existence of both near-peer and irregular warfare threats requiring airborne electronic attack solutions. The Army and Marine Corps requested that the analysis address irregular warfare threats because of the growing concern over improvised explosive devices in Iraq and Afghanistan and the suboptimal application of existing systems in the inventory to defeat those threats. The Air Force concluded in its analysis that fulfilling airborne electronic attack mission needs would require developing and fielding multiple new systems.

The most recent study, U.S. Strategic Command's Electronic Warfare Initial Capabilities Document, identified additional capability gaps affecting airborne electronic attack. This 2009 analysis built upon a capabilities-based assessment completed a year earlier and outlined mitigation strategies to address these gaps instead of merely prescribing specific platform solutions. This approach was consistent with the analysis's charter to guide and inform the services' acquisition programs. However, the analysis did recommend specific capabilities and system attributes for the Next Generation Jammer program to consider that would assist in mitigating some of the gaps identified in the 2009 analysis. The analysis also concluded that new systems would be needed to close nearly half of the gaps identified in airborne electronic attack capabilities.

Improvements to Tactics, Techniques, and Procedures and Investments in Science and Technology Are Helping to Bridge Gaps

To supplement its acquisition of new systems, DOD is undertaking other efforts to bridge existing airborne electronic attack capability gaps. In the near term, services are evolving their tactics, techniques, and procedures for operating existing systems to enable them to take on additional mission tasks. These activities maximize the utility of existing systems and better position operators to complete missions with equipment currently available. Longer-term solutions, however, depend on the department successfully capitalizing on its investments in science and technology. DOD has recently taken actions that begin to address long-standing coordination shortfalls in this area including designating electronic warfare as a priority area for investment and creating a steering council to link capability gaps to research initiatives. However, these steps do not preclude services from funding their own research priorities ahead of departmentwide priorities. DOD's planned implementation roadmap for electronic warfare offers an opportunity to assess how closely component research investments are aligned to the departmentwide electronic warfare priority.

Changing Tactics, Techniques, and Procedures for Existing Systems Can Mitigate Gaps in the Near Term

The refinement of tactics, techniques, and procedures can position the services to maximize the capabilities of existing systems while new capabilities are being developed. As Navy airborne electronic attack operators stated, when a capability gap requiring a new system is identified, warfighters generally do not have the luxury of waiting for the acquisition community to develop and field a system to fill that gap. In the interim, tactics, techniques, and procedures for existing systems must evolve to provide at least partial mitigation to the threat being faced. Development and refinement of new ways to use existing equipment allow the services to maximize the utility of their airborne electronic attack systems and leave them better positioned to complete missions with the assets they have available. The following two systems provide examples where operator communities have refined tactics, techniques, and procedures to meet emerging threats:

- **AN/ALQ-99 Tactical Jamming System:** Navy officials told us that threats encountered in Iraq and Afghanistan operations have driven significant changes to how the AN/ALQ-99 Tactical Jamming System is employed. In essence, tactics, techniques, and procedures for the system had to evolve to maximize the system's capabilities against irregular warfare threats. According to Navy officials, however, these adaptations represent only a temporary solution as their application—coupled with increased operational activity—has caused jamming pods to degrade and burn out at an increasing rate, subsequently increasing maintenance requirements for the system.
- **EC-130H Compass Call:** According to Air Force officials, EC-130H tactics, techniques, and procedures have rapidly evolved to encompass dynamically changing electronic attack threats, which include

irregular warfare. These changes include modifications to both how the operator employs the aircraft as well as to the range of threats targeted by mission planners.

Both Navy and Air Force officials emphasized that sustained investments in tactics, techniques, and procedures offer considerable return on investment and can provide important, near-term solutions to longer-term, persistent threats. According to these officials, these investments position operators to "do more with less"—in effect, offer them the opportunity to mitigate or counteract a threat without the required new system. However, limits exist to the extent to which refinements to current operating approaches for existing systems can bridge capability gaps. For example, it is increasingly difficult to further optimize AN/ALQ-99 jamming pods to counter advanced, integrated air defense systems. Specifically, Navy officials stated that the AN/ALQ-99 has reached its limit in terms of the underlying architecture's capability to grow to counter new, sophisticated types of threats.

DOD Focusing Science and Technology Investments to Close Gaps in the Long Term, but Coordination Remains a Concern

Investment in the science and technology research base is a longer-term approach DOD uses to address capability gaps in mission areas. Electronic warfare, including airborne electronic attack, is supported by research investments in fields such as sensors, apertures, power amplifiers, and unmanned aircraft technology that may help address existing capability gaps. Service components categorize research investments differently from one another, which complicates efforts to clearly define funding devoted to airborne electronic attack. Table 6 identifies some of DOD's current airborne electronic attack-related research investments.

However, not all investments in these fields will necessarily improve airborne electronic attack capabilities. Research officials identify the transition to system development and procurement as one of the primary goals of defense research programs, but acknowledge, reasonably, that not every program will successfully develop a transitionable product. Some acquisition programs, such as the Next Generation Jammer and the MQ-9 Reaper Electronic Attack Pod, invest directly in research to guide the transition process and increase the likelihood of success. But even with this direct attention, technology maturation and development for Next Generation

Jammer is expected to last 8 to 9 years. Consequently, current science and technology initiatives represent a long-term investment in future capabilities and are less suited to meeting existing needs.

DOD analyses during the past decade have identified coordination deficiencies that constrain the department's ability to capitalize on its science and technology investments. For instance, a 2005 Naval Research Advisory Committee report found that within the Navy, research and development efforts were unduly fragmented, with one laboratory or development activity often being unaware of what another was doing.[15] Further, this study highlighted the lack of a long-range science and technology investment planning process within the Navy. Similarly, in 2007, the Defense Science Board reported that although relevant and valuable science and technology activity was occurring, an overarching, departmentwide strategic technology plan with assigned responsibility, accountability, and metrics did not exist.[16] According to the board, DOD's science and technology activities and investments should be more directly informed by the department's strategic goals and top-level missions—an objective that would require a closer coupling of technologists and users, including requirements and capabilities developers. A 2010 Naval Research Advisory Committee report[17] built on previous findings noted that stewardship of long-term naval capabilities was "vague at best" and lacked specific organizational assignment.[18] The report recognized the Navy as having the lead role within DOD for electronic warfare, but identified sporadic and uncoordinated execution across the technical community—noting little evidence of engagement among the science and technology community at large. Further, the report advised that closer coordination between operational and technical communities was essential for the realization of desired long-term capabilities.

DOD has recently taken actions that begin to address these shortfalls, including formalizing existing investment processes for several key science and technology areas. Most notably, in April 2011 the Secretary of Defense designated electronic warfare as one of seven priority areas for science and technology investment from fiscal years 2013 through 2017. According to officials from the Office of the Assistant Secretary of Defense for Research and Engineering (ASD(R&E)), this designation carries the promise of increased research funding and has prompted chartering of the interdepartmental Electronic Warfare Priority Steering Council. This council is made up of research officials from ASD (R&E), the services, and various defense science and technology groups, such as the Defense Advanced Research Projects Agency, and is charged with effectively evaluating electronic warfare capability gaps and linking them with research initiatives necessary to fill them. To support this process, the council is developing an implementation roadmap to guide coordination of investments within the electronic warfare area. The council also facilitates ASD(R&E) coordination with requirements teams and service/external research offices to determine the specific fields of inquiry that will be needed to support planning for future electronic warfare capability needs. Previously, this coordination was handled informally, whereas the new council provides authority and visibility to the discussions and decisions made.

Table 6. Current DOD Science and Technology Initiatives Related to Airborne Electronic Attack

Agency	Acquisition vehicle	Examples of funded programs/fields	Budgeted funds
Office of Naval Research	Long-range broad agency announcement	Electronics, Sensors, & Network Research Receivers & Antennas Power Amplifiers	A total of $4.0 million in fiscal year 2011 for all electronic warfare research, but new plans are to increase this amount to approximately $24.0 million annually
Air Force Research Laboratory	Research interests of the office's broad agency announcement	Electro-Energetic Physics Materials & Metamaterials Receiver Technology	A total of $29.7 million in fiscal year 2011 for all electronic warfare research, decreasing to $24.2 million in fiscal year 2012, with plans to increase funding in fiscal year 2013
Defense Advanced Research Projects Agency (DARPA)	Project-specific broad agency announcements	Behavioral Learning for Adaptive Electronic Warfare Precision Electronic Warfare	A total of $20.7 million in fiscal year 2011 and $18.8 million in fiscal year 2012 for these two electronic warfare research programs

Source: GAO analysis of DOD data.

Notwithstanding these important steps, services may inevitably face situations where they have to choose between funding their own, service-specific research priorities and funding departmentwide priorities. As the Assistant Secretary of Defense for Research and Engineering testified in 2011, DOD's seven priority areas for science and technology investment are meant to be in addition to the priorities outlined by individual components (i.e., service research agencies and DARPA).[19] In other words, departmentwide science and technology priorities do not necessarily supplant service priorities. Absent strategic direction, however, services have generally been inclined to pursue their own research interests ahead of departmentwide pursuits. DOD's planned implementation roadmap for electronic warfare offers opportunities to assess how closely component research investments are aligned to the departmentwide electronic warfare priority and to coordinate component investments in electronic warfare.

CONCLUSION

The rapidity of evolving threats, together with the time and cost associated with fielding new systems, creates a major challenge to DOD and its capacity to fill all of its capability gaps. This dynamic makes it imperative that the department get the most out of its electronic warfare investments. At this point, that does not appear to be the case. The systems being acquired have problems and will not deliver as expected; potential overlap, to the extent that it leads to covering some gaps multiple ways while leaving others uncovered, drains buying power from the money that is available; and DOD acknowledges a leadership void that makes it difficult to ascertain whether the current level of investment is optimally matched with the existing capability gaps.

Within the airborne electronic attack mission area, budgetary pressures and related program cancellations prompted the department to change its acquisition strategy from a system of systems construct—as underpinned by the 2002 analysis of alternatives—to a potentially less robust, but more affordable, family of systems. In addition, new systems, including AARGM and MALD, that are designed to replace or augment legacy assets have encountered technical challenges while in acquisition, subsequently requiring the services to delay fielding plans within each program. Other acquisition programs, including IDECM and MALD-J, are structured with a high degree of concurrency between development, production, and testing that position them for similar suboptimal outcomes. Although individual service decisions to delay or cancel underperforming or resource-intensive programs may be fiscally prudent, the cumulative effect of these decisions creates uncertainty as to when, or if, current departmentwide airborne electronic attack capability gaps can be filled. At present, even if the department successfully acquires the full complement of systems outlined in its family of systems strategy, some capability gaps identified a decade ago may persist. As such, the department can benefit from reevaluating its capability gaps—using structures like the new Electronic Warfare Priority Steering Council—to identify which ones are highest priorities for science and technology investment and to determine areas where it is more willing to accept mission risk. This analysis, when coupled with an examination of current service-specific science and technology investments, can position DOD to realize improved efficiencies in its electronic warfare research activities and better align constrained budgets with highest-priority needs. Additionally, because underperformance in acquisition programs exacerbates existing capability gaps, realistic assessments of higher-risk programs can provide needed insight into what capabilities each platform is likely to deliver and when. Shortfalls in acquisition should not be the deciding factor on which capability gaps the department accepts.

At the same time, services continue to pursue and invest in multiple separate airborne electronic attack systems that potentially overlap with one another. This overlap is most evident in irregular warfare systems, including the Marine Corps's Intrepid Tiger II and the Army's CEASAR systems, but is also present in Air Force and Navy efforts to develop expendable jamming decoys through their respective MALD-J and Airborne Electronic Attack Expendable initiatives. Pursuing multiple separate acquisition efforts to develop similar capabilities can result in the same capability gap being filled twice or more, can lead to inefficient use of resources, and may contribute to other warfighting needs going unfilled. Leveraging resources and acquisition efforts across services—not just by sharing information, but through shared partnerships and investments—can simplify developmental efforts, can improve interoperability among systems and combat forces, and could decrease future operating and support costs. Such successful outcomes can position the department to maximize the returns it gets on its airborne electronic attack investments.

Recommendations for Executive Action

We recommend that the Secretary of Defense take the following five actions:

- Given airborne electronic attack programmatic and threat changes since 2002, complete the following:

  - Conduct program reviews for the AARGM, IDECM, MALD, and MALD-J systems to assess cost, schedule, and performance and direct changes within these investments, as necessary.
  - Determine the extent to which the most pressing airborne electronic attack capability gaps can best be met—using the assets that are likely to be available—and take steps to fill any potential gaps.
  - Align service investments in science and technology with the departmentwide electronic warfare priority, recognizing that budget realities will likely require trade-offs among research areas, and direct changes, as necessary.
- To ensure that investments in airborne electronic attack systems are cost-effective and to prevent unnecessary overlap, take the following actions:
  - Review the capabilities provided by the Marine Corps's Intrepid Tiger II and Army's CEASAR systems and identify opportunities for consolidating these efforts, as appropriate.
  - Assess Air Force and Navy plans for developing and acquiring new expendable jamming decoys, specifically those services' respective MALD-J and Airborne Electronic Attack Expendable initiatives, to determine if these activities should be merged.

AGENCY COMMENTS AND OUR EVALUATION

We provided a draft of this report to DOD for comment. In its written comments, DOD concurred with three of our recommendations and partially concurred with two recommendations. DOD also provided technical comments that we incorporated into the report, as appropriate.

DOD concurred with our first recommendation to conduct program reviews for the AARGM, IDECM, MALD, and MALD-J systems and direct changes within these investments, as necessary, identifying a March 2012 Navy review of the IDECM program and planned July 2012 Navy review of the AARGM system. For MALD and MALD-J, DOD plans to conduct a program review in early 2014, which will coincide with a planned full rate production decision for MALD-J. In the interim, DOD intends to continue low rate initial production of MALD-J units. However, because MALD has experienced significant technical challenges within the past 2 years, and because DOD plans to invest an additional $176.9 million toward MALD-J production through fiscal year 2014, we believe an earlier review may be warranted. In its written comments, DOD also stated that the Deputy Assistant Secretary of Defense for Strategic and Tactical Systems will chair a meeting to review AARGM, IDECM, MALD, and MALD-J with the Navy and Air Force to verify progress, but it did not provide a timetable for this review.

DOD also concurred with our second recommendation to determine the extent to which the most pressing airborne electronic attack capability gaps can best be met—using the assets that are likely to be available—and take steps to fill any potential gaps. Most notably, DOD cited plans for U.S. Strategic Command to annually assess all DOD electronic warfare capabilities—including current requirements, current and planned future capabilities, and the supporting investment strategy—and present this assessment to the Joint Requirements Oversight Council. Further, DOD concurred with our third recommendation to align service investments in science and technology with the departmentwide electronic warfare priority, noting in its written comments that it expects implementation roadmaps for priority areas (including electronic warfare) will serve to coordinate component investments and accelerate the development and delivery of capabilities.

DOD partially concurred with our two recommendations related to potentially unnecessary overlap among airborne electronic attack systems, identifying through its written comments plans for the Deputy Assistant Secretary of Defense for Strategic and Tactical Systems to review the Intrepid Tiger and CEASAR systems with the Marine Corps and Army to investigate the efficacy of additional coordination as future acquisition plans are evaluated. Similarly, DOD noted that following the expected March 30, 2012, completion of a new Air Force plan related to developing and procuring an Increment II variant of MALD-J, the Office of the Under Secretary of Defense for Acquisition, Technology and Logistics; Office of the Director, Cost Assessment and Program Evaluation; and Joint Staff would review Air Force and Navy plans and assess opportunities for coordination among the MALD-J and Airborne Electronic Attack Expendable initiatives, should funding be allocated for a future expendables program. However, the basis for DOD's partial agreement on these two recommendations appears to stem from its desire to achieve efficiencies through increased coordination among programs—not through consolidation of systems possessing similar capabilities. We emphasize that coordination is not a substitute for consolidation—particularly in the current constrained budget environment—and we encourage DOD to expand the scope of its planned reviews to include assessments of potential unnecessary redundancies within these two sets of systems.

Additionally, DOD commented that our draft report overstated the acquisition duplication among airborne electronic attack systems. Most notably, DOD pointed to its cancellations of the MQ-9 Electronic Attack Pod and MALD-J Increment II programs, as outlined in its fiscal year 2013 budget submission, as evidence that duplication was being managed. These cancellations were announced after we had completed our work and drafted the report. During the period that our draft report was with the agency for comment, we revised our report and recommendations, in coordination with DOD, to account for these recent changes. Most notably, we revised our fourth and fifth recommendations to remove the newly canceled MQ-9 Electronic Attack Pod and MALD-J Increment II systems, respectively, as additional platforms where DOD may identify opportunities for consolidation. DOD's written comments were subsequently crafted in response to our revised set of recommendations. As noted above, opportunities to reduce duplication further remain. We also briefly introduced the Marine Air Ground Task Force Electronic Warfare concept, in response to DOD's comments, while further clarifying that our report did not evaluate ground- or ship-based electronic warfare systems.

DOD also commented that our characterization of the family of systems strategy for airborne electronic attack was misleading, stating that the system of systems synergies envisioned in 2002 continue to be pursued. We acknowledge that DOD is considering options to field additional systems against high-end threats, but we believe that the current acquisition strategy and its distributed approach is very much in line with the definition of a family of systems, as outlined by DOD.[20] When DOD embarked on the system of systems strategy in 2002, it envisioned fielding certain major systems, such as B-52 Standoff Jammer and J-UCAS, which were later canceled. Without these planned elements, there is no evidence to suggest that the remaining systems together possess capability beyond the additive sum of the individual capabilities of its members—a characteristic fundamental to a system of systems.

Michael J. Sullivan
Director
Acquisition and Sourcing Management

APPENDIX I. SCOPE AND METHODOLOGY

This report evaluates the Department of Defense's (DOD) airborne electronic attack capabilities and investment plans.[21] Specifically, we assessed (1) the department's strategy for acquiring airborne electronic attack capabilities, (2) progress made developing and fielding systems to meet airborne electronic attack mission requirements, and (3) additional compensating actions taken by the department to address capability gaps, including improvements to tactics, techniques, and procedures and investments in science and technology. To assess the department's strategy for acquiring airborne electronic attack capabilities, we analyzed DOD's documents outlining mission requirements and acquisition needs, including the 2002 Airborne Electronic Attack Analysis of Alternatives, 2004 Initial Capabilities Document for Denying Enemy Awareness through Airborne Electronic Attack, 2008 Electronic Warfare Capabilities-Based Assessment, 2009 Electronic Warfare Initial Capabilities Document, and 2010 Electronic Warfare Strategy of the Department of Defense report to Congress. We also reviewed platform-specific capabilities documents, service roadmaps related to airborne electronic attack, and budget documents to understand how the family of systems construct evolved over time. To identify capability limitations and sustainment challenges facing current airborne electronic attack systems, we reviewed program briefings and acquisition documentation related to these systems. To further corroborate documentary evidence and obtain additional information in support of our review, we conducted interviews with relevant DOD officials responsible for managing airborne electronic attack requirements and overseeing the related family of systems, including officials in the Office of the Under Secretary of Defense for Acquisition, Technology and Logistics; Office of the Director, Cost Assessment and Program Evaluation; Office of the Assistant Secretary of the Navy for Research, Development and Acquisition; Office of the Chief of Naval Operations—Information Dominance and Air Warfare directorates; Office of the Assistant Secretary of the Air Force for Acquisition; Air Force Office of the Deputy Chief of Staff for Operations, Plans, and Requirements—Electronic Warfare division; Air Force Air Combat Command; Army Office of the Deputy Chief of Staff for Operations, Plans, and Training—Electronic Warfare division; Marine Air-Ground Task Force Electronic Warfare; U.S. Strategic Command; and Joint Staff. We also held discussions with DOD officials responsible for sustaining current airborne electronic attack systems, including officials in (1) Navy program offices for Airborne Electronic Attack, Advanced Tactical

Aircraft Protection Systems, Direct and Time Sensitive Strike, and Aerial Target and Decoy Systems and (2) Air Force offices, including the F-22A Raptor and F-16CM program offices and Warner Robins Air Logistics Center.

To assess progress made developing and fielding systems to meet airborne electronic attack mission requirements, we analyzed documents outlining acquisition plans, costs, and performance outcomes, including capabilities documents, program schedules, test reports, budget submissions, system acquisition reports, and program briefings. These same materials afforded information on key attributes of individual airborne electronic attack systems, which we used to assess potential overlap among systems in development. Further, we identified persisting airborne electronic attack capability gaps by reviewing the 2009 Electronic Warfare Initial Capabilities Document, along with earlier analyses related to airborne electronic attack requirements, and compared the capability needs identified in those documents with current DOD investments in airborne electronic attack capabilities. To supplement our analyses and gain additional visibility into these issues, we conducted interviews with relevant DOD officials responsible for managing airborne electronic attack requirements, including officials in the Office of the Chief of Naval Operations—Information Dominance and Air Warfare directorates; Office of the Assistant Secretary of the Air Force for Acquisition; Air Force Office of the Deputy Chief of Staff for Operations, Plans, and Requirements Electronic Warfare division; Air Force Air Combat Command; Army Office of the Deputy Chief of Staff for Operations, Plans, and Training—Electronic Warfare division; Marine Air-Ground Task Force Electronic Warfare; U.S. Strategic Command; and Joint Staff. We also held numerous interviews with DOD officials primarily responsible for developing, acquiring, and testing airborne electronic attack systems, including officials in the Office of the Under Secretary of Defense for Acquisition, Technology and Logistics; Office of the Director, Operational Test and Evaluation; Office of the Deputy Assistant Secretary of Defense for Developmental Test and Evaluation; Office of the Assistant Secretary of the Navy for Research, Development and Acquisition; Office of the Assistant Secretary of the Air Force for Acquisition; Navy program offices for Airborne Electronic Attack, F/A-18 and EA-18G, Direct and Time Sensitive Strike, and Advanced Tactical Aircraft Protection Systems; Army Rapid Equipping Force; and Air Force program offices for MALD/MALD-J and MQ-9 Reaper Electronic Attack Pod. To assess additional compensating actions taken by the department to address airborne electronic attack capability gaps, we reviewed service documents outlining recent improvements and refinements to tactics, techniques, and procedures for EA-18G and EC-130H aircraft. We corroborated this information through interviews with officials from the Naval Strike and Air Warfare Center and Air Force Office of the Deputy Chief of Staff for Operations, Plans, and Requirements—Electronic Warfare division charged with refining tactics, techniques, and procedures for EA-18G and EC-130H aircraft. We also reviewed broad agency announcements to understand ongoing science and technology activities related to airborne electronic attack. We supplemented this documentation review with discussions with officials engaged in science and technology work tied to airborne electronic attack, including officials in the Office of the Assistant Secretary of Defense for Research and Engineering, Office of Naval Research, Air Force Research Laboratory, and Defense Advanced Research Projects Agency.

We conducted this performance audit from February 2011 to March 2012 in accordance with generally accepted government auditing standards. Those standards require that we plan and perform the audit to obtain sufficient, appropriate evidence to provide a reasonable basis for our findings and conclusions based on our audit objectives. We believe that the evidence obtained provides a reasonable basis for our findings and conclusions based on our audit objectives.

APPENDIX II. ANALYSES OF SELECT AIRBORNE ELECTRONIC ATTACK SYSTEMS

This appendix provides analyses of 10 selected airborne electronic attack systems. Figures 4 through 13 show images of each system; tables 7 through 16 provide budget data on each system.

Source: Northrop Grumman.

Figure 4. EA-6B Prowler.

Estimated end of service life: 2020

Mission description: The primary mission of the Prowler is the suppression of enemy air defenses in support of strike aircraft and ground troops by interrupting enemy electronic activity and obtaining tactical electronic intelligence within the combat area. The EA-6B uses the AN/ALQ-99 radar jamming pod for non-lethal protection by jamming air defense systems and its AGM-88 High Speed Anti-Radiation Missile for lethal physical attack of air defense systems.

Status: In 2010, we reported that the Navy had started replacing its EA-6B aircraft with EA-18G Growlers and expected all Prowlers to be out of its inventory by 2012. However, the Navy projects Prowlers to remain in service until 2016 to further meet the joint expeditionary need. According to the Navy, this is subject to additional change contingent on the fiscal year 2013 budget. The Marine Corps plans to retire its Prowlers by 2020. In addition, the most recent upgrade program for the EA-6B—the third Improved Capability electronic suite modification (ICAP III)—is nearing completion. ICAP III provides the Prowler with greater jamming capability, including the ability to perform selective reactive jamming.

Budget: See the following table for budget information.

Table 7. DOD Planned Acquisition Investments for the EA-6B Prowler, Fiscal Years 2012-2017

Then-year dollars in millions							
	FY 2012	**FY 2013**	**FY 2014**	**FY 2015**	**FY 2016**	**FY 2017**	**Total**
RDT&E	$20.222	$19.728	$19.931	$20.280	$20.252	$20.632	**$121.045**
Procurement	27.734	30.062	18.600	14.099	10.068	10.285	**$110.848**
Total	**$47.956**	**$49.790**	**$38.531**	**$34.379**	**$30.320**	**$30.917**	$231.893

Legend: RDT&E = research, development, testing, and evaluation.
Source: Department of the Navy fiscal year 2013 budget estimates.
Note: RDT&E funding is limited to electronic warfare counter response.

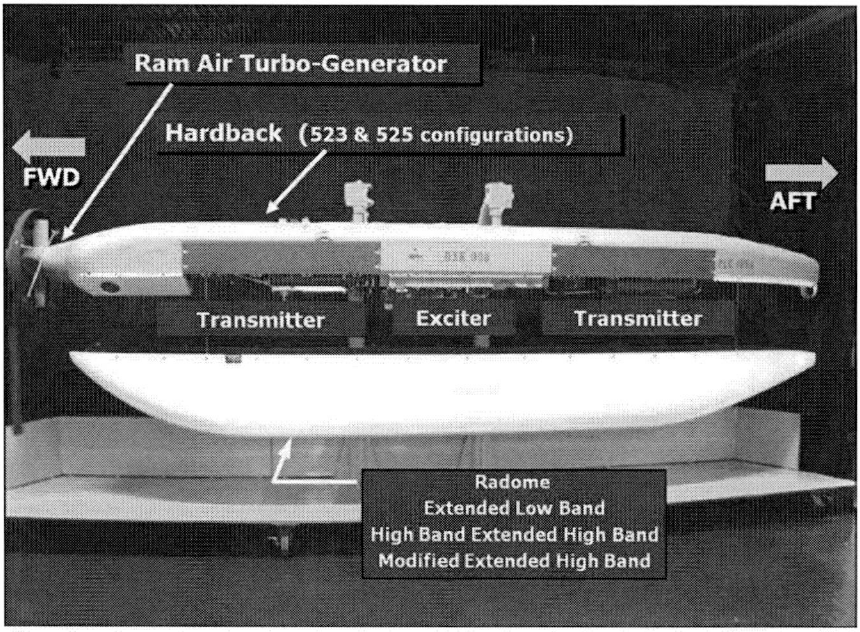

Source: NAVAIR PMA-234.

Figure 5. AN/ALQ-99 Tactical Jamming System.

Estimated end of service life:
Mid-band: 2024
Low-band: 2026
High-band: 2028

Mission description: The AN/ALQ-99 Tactical Jamming System is an airborne electronic warfare system carried on the EA-6B and EA-18G to support the suppression of enemy air defenses. The system is capable of intercepting, automatically processing, and jamming received radio frequency signals.

Status: Obsolescence issues and advances in adversary technology have reduced the AN/ALQ-99's ability to counter emerging threats. The Navy is developing its Next Generation Jammer program to replace the AN/ALQ-99 and plans to begin fielding the system in 2020.

In the interim, the Navy is currently replacing three aging legacy low-band transmitters to resolve obsolescence and reliability problems.

Budget: See the following table for budget information.

Table 8. DOD Planned Acquisition Investments for the AN/ALQ-99 Tactical Jamming System, Fiscal Years 2012-2017

Then-year dollars in millions							
	FY 2012	**FY 2013**	**FY 2014**	**FY 2015**	**FY 2016**	**FY 2017**	**Total**
RDT&E	-	-	-	-	-	-	-
Procurement	$69.665	$49.799	$40.078	$28.892	$35.963	$30.945	**$255.342**
Total	**$69.665**	**$49.799**	**$40.078**	**$28.892**	**$35.963**	**$30.945**	**$255.342**

Legend: RDT&E = research, development, testing, and evaluation.

Source: Department of the Navy fiscal year 2013 budget estimates.

Note: There is no RDT&E funding associated with the AN/ALQ-99 Tactical Jamming System in the fiscal year 2013 budget.

Source: 755 Maintenance Squadron, U.S. Air Force.

Figure 6. EC-130H Compass Call.

Estimated end of service life: 2053

Mission description: The EC-130H Compass Call is an airborne, wide area, persistent stand-off electronic attack weapon system able to disrupt and deny adversary use of the electronic battlespace using offensive radio frequency countermeasures. Its primary mission is to deny or disrupt command and control of enemy integrated air defenses, air defense surface-to-air missile and anti-aircraft artillery threats. Its secondary mission is to support ground and special operations forces by denying enemy communications and defeating improvised explosive devices.

Status: The Air Force has evolved the Compass Call since it was first fielded in 1982 to meet modern and emerging threats, including commercial communications, early warning radars, and improvised explosive devices. Upgrades and modernization efforts are completed during regularly scheduled depot maintenance.

In 2003, as a response to Operation Enduring Freedom, these upgrades transitioned from "Block" upgrades to "Baseline" upgrades to allow for smaller and more focused modernization efforts. Currently, the Air Force is completing Baseline 1 upgrades, beginning Baseline 2 efforts, and developing Baseline 3 requirements.

In addition, the Air Force is also replacing the center wing box on all 14 Compass Call aircraft, which will extend the service life of the fleet. Compass Call has been on continuous deployment in support of operations in Iraq and Afghanistan since 2003; which has accelerated the need to replace the center wing boxes. Finally, to further alleviate stress on the fleet, the Air Force plans to procure an additional aircraft, increasing the size of the fleet to 15 aircraft by fiscal year 2016.

Budget: See the following table for budget information.

Table 9. DOD Planned Acquisition Investments for the EC-130H Compass Call, Fiscal Years 2012-2017

Then-year dollars in millions

	FY 2012	FY 2013	FY 2014	FY 2015	FY 2016	FY 2017	Total
RDT&E	$18.509	$12.094	$12.222	$12.559	$13.047	$12.989	**$81.420**
Procurement	302.324	64.024	55.878	54.108	56.480	57.552	**$590.366**
Total	**$320.833**	**$76.118**	**$68.100**	**$66.667**	**$69.527**	**$70.541**	**$671.786**

Legend: RDT&E = research, development, testing, and evaluation.
Source: Department of the Air Force fiscal year 2013 budget estimates.

Source: U.S. Air Force.

Figure 7. F-22A Raptor.

Estimated end of service life: Not available

Mission description: The F-22A is the Air Force's fifth-generation air superiority fighter that incorporates a stealthy and highly maneuverable airframe, advanced integrated avionics, and a supercruise engine. Originally developed as an air-to-air fighter, additional capabilities will allow the F-22A to perform multiple missions including destruction of enemy air defenses, air-to-ground attack, electronic attack, and intelligence surveillance and reconnaissance.

Status: The F-22A, along with the F-35, is expected to fulfill the Air Force's requirement for penetrating escort jamming capability. The Air Force initiated a formal F-22A modernization and reliability improvement program in 2003 to incrementally develop and deliver increasing capabilities over time. These increasing capabilities would allow the F-22A to provide penetrating escort jamming, as envisioned in the airborne electronic attack family of systems strategy. However, fielding of these capabilities has been delayed because of reductions in program funding. In addition, we have previously reported on schedule delays within the modernization and reliability improvement program and their effect on fielding additional capabilities within expected time frames. Further delays in fielding these planned capabilities may affect the Air Force's ability to provide sufficient penetrating escort jamming, increasing mission risk.

Budget: See the following table for budget information.

Table 10. DOD Planned Acquisition Investments for the F-22A Raptor, Fiscal Years 2012-2017

Then-year dollars in millions							
	FY 2012	**FY 2013**	**FY 2014**	**FY 2015**	**FY 2016**	**FY 2017**	**Total**
RDT&E	$571.320	$511.767	$503.242	$387.510	$430.947	$463.263	**$2,868.049**
Procurement	232.032	283.871	291.741	248.001	282.249	329.775	**$1,667.669**
Total	**$803.352**	**$795.638**	**$794.983**	**$635.511**	**$713.196**	**$793.038**	**$4,535.718**

Legend: RDT&E = research, development, testing, and evaluation.

Source: Department of the Air Force fiscal year 2013 budget estimates.

Note: The above budget figures are only for F-22A modernization efforts only and do not include $104.118 million in fiscal year 2012 funds for equipment, program support, and shutdown activities necessary to preserve assets for long-term F-22A fleet sustainment.

Source: Boeing.

Figure 8. EA-18G Growler.

Fielding date: 2009

Mission description: The EA-18G Growler replaces the EA-6B Prowler as DOD's tactical electronic attack aircraft. Like the Prowler, the EA-18G will provide full-spectrum electronic attack to counter enemy air defenses and communication networks. The EA-18G incorporates jamming capabilities, such as the AN/ALQ-99 Tactical Jamming System, and the use of onboard weapons such as the High Speed Anti-Radiation Missile, for the suppression of enemy air defenses. The Growler is the Navy's platform to fulfill modified escort jamming capability needs.

Status: The Growler program entered full rate production in 2009, with a planned acquisition of 88 aircraft. However, in 2009, the Office of the Secretary of Defense directed the Navy to buy an additional 26 aircraft, bringing the total units to be acquired to 114. Through fiscal year 2011, the Navy placed 90 of 114 planned EA-18G aircraft under contract for production. Production is slightly ahead of schedule and has incorporated the increase in total units with limited per-unit cost growth.

In 2010, the Director, Operational Test and Evaluation, declared the Growler operationally effective, but also found that the aircraft was unsuitable for operations based on maintainability concerns. Since then, the Navy has taken steps to improve the EA-18Gs suitability through software fixes, and the system recently completed follow-on operational test and evaluation. In addition, initial deployment of the aircraft in support of operations in Iraq, Libya, and Afghanistan recently concluded, and the Navy is assessing the aircraft's performance, including the remaining challenges mitigating electromagnetic interference with the AN/ALQ-99. Additional software improvements are planned through fiscal year 2018.

Budget: See the following table for budget information.

Table 11. DOD Planned Acquisition Investments for the EA-18G Growler, Fiscal Years 2012-2017

Then-year dollars in millions							
	FY 2012	**FY 2013**	**FY 2014**	**FY 2015**	**FY 2016**	**FY 2017**	**Total**
RDT&E	$17.100	$13.009	$15.311	$16.002	$16.106	$16.393	**$93.921**
Procurement	1,022.715	1,027.443	21.970	8.111	0.000	0.000	**$2080.239**
Total	**$1,039.815**	**$1,040.452**	**$37.281**	**$24.113**	**$16.106**	**$16.393**	**$2,174.160**

Legend: RDT&E = research, development, testing, and evaluation.

Source: Department of the Navy fiscal year 2013 budget estimates.

Note: The above budget figures do not include $34.151 million in fiscal year 2013 for procurement of initial spares.

Source: NAVAIR PMA-242.

Figure 9. AGM-88E Advanced Anti-Radiation Guided Missile (AARGM).

Estimated fielding date: 2012

Mission description: AARGM is an air-to-ground missile for carrier-based aircraft designed to destroy enemy radio-frequency-enabled surface-to-air defense. AARGM is an upgrade to the AGM-88 High Speed Anti-Radiation Missile (HARM) and will utilize existing HARM propulsion and warhead sections with new guidance and modified control sections.

Status: The Navy authorized AARGM production in September 2008, with deliveries scheduled to begin in January 2010. A total of 1,919 units were planned, with initial operational capability scheduled for March 2011. The program began operational testing in June 2010 after a 9-month delay owing, in part, to concerns about the production representativeness of test missiles. The Navy halted operational testing in September 2010 after hardware and software deficiencies caused a series of missile failures.

These testing challenges prompted the Navy to delay AARGM's planned initial operational capability date and undertake corrective actions to the system. These actions included an evaluation of the AARGM system through laboratory, ground, and flight tests from November 2010 through June 2011. Following this testing, Navy officials concluded that previous testing anomalies were successfully corrected but that the system was at high risk of not meeting suitability requirements during operational testing. The Navy found that insufficient system reliability and manufacturing quality controls remain open deficiencies that will likely result in an excessive number of system failures experienced by operational units, which could prevent the Navy from effectively executing planned missions. To address reliability concerns, the Navy instituted a "fly before you buy" program to screen poor weapons prior to government acceptance. As of July 2011, one-third of missiles delivered for testing were returned to the factory for repair.

Recently, the AARGM system resumed operational testing. The Navy now plans to field the system beginning in April 2012 and make a full rate production decision and contract award in June and July 2012, respectively.

Budget: See the following table for budget information.

Table 12. DOD Planned Acquisition Investments for AARGM, Fiscal Years 2012-2017

Then-year dollars in millions							
	FY 2012	**FY 2013**	**FY 2014**	**FY 2015**	**FY 2016**	**FY 2017**	**Total**
RDT&E	$6.684	$6.995	$7.426	$5.470	$5.142	$5.028	**$36.745**
Procurement	71.561	86.721	112.022	126.324	158.073	160.820	**$715.521**
Total	**$78.245**	**$93.716**	**$119.448**	**$131.794**	**$163.215**	**$165.848**	**$752.266**

Legend: RDT&E = research, development, testing, and evaluation.

Source: Department of the Navy fiscal year 2013 budget estimates.

Note: The above budget figures do not include $0.209 million in fiscal year 2012 for procurement of initial spares.

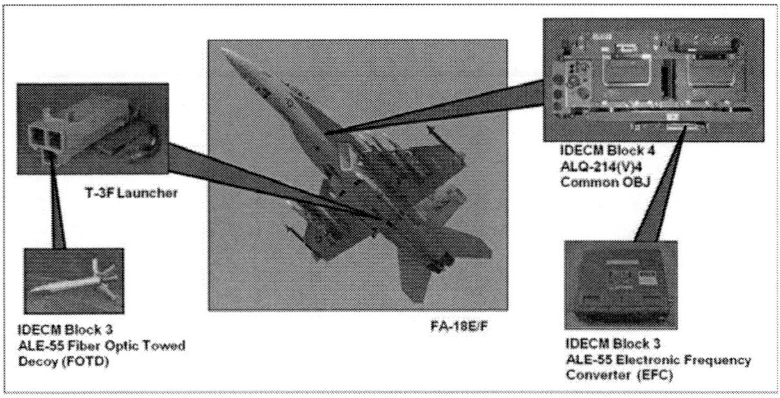

Source: U.S. Navy.

Figure 10. Integrated Defensive Electronic Countermeasures (IDECM).

Estimated fielding date: 2014 (Block 4)

Mission description: IDECM is a suite of self-protection countermeasure systems designed for the F/A-18E/F, including onboard jamming and off-board decoy jamming capabilities. The Navy has fielded IDECM in different blocks dating back to 2002 (Block 1), 2004 (Block 2), and 2011 (Block 3). Each block improved the system's jamming capabilities, decoy capabilities, or both. Block 4—the phase of production currently in development—extends IDECM onboard jamming capabilities to F/A-18C/D aircraft.[22]

Status: IDECM Block 4 entered development in 2009 and includes redesign of the ALQ-214 onboard jammer from the component design used for earlier blocks. This redesign is driven by the need to reduce weight in order to accommodate the IDECM onboard system on F/A-18C/D aircraft. Essentially, the new ALQ-214 will perform the same onboard jammer function as found in IDECM Blocks 2 and 3 but with a different form and fit. The Navy expects to transition current IDECM Block 3 full rate production to Block 4 units by April 2012. This production transition will occur concurrent with ground and flight testing of the Block 4 system—a strategy that could drive costly design changes, retrofits, or both to units in production, in the event that the ALQ-214 redesign effort does not materialize on schedule. To mitigate this risk, Navy officials stated that Block 4 full rate production will initially be for 19 systems, with production rates increasing to as many as 40 per year following completion of testing. Further, DOD officials report that Block 4 production will be executed under a firm fixed-price contract—a strategy that DOD officials state will place the financial burden of any retrofits on the vendor.

Budget: See the following table for budget information.

Table 13. DOD Planned Acquisition Investments for IDECM, Fiscal Years 2012-2017

Then-year dollars in millions							
	FY 2012	**FY 2013**	**FY 2014**	**FY 2015**	**FY 2016**	**FY 2017**	**Total**
RDT&E	$62.100	$29.874	$14.408	$13.897	$2.711	$2.848	**$125.838**
Procurement	40.272	57.067	84.305	102.388	133.449	51.569	**$469.050**
Total	**$102.372**	**$86.941**	**$98.713**	**$116.285**	**$136.160**	**$54.417**	**$594.888**

Legend: RDT&E = research, development, testing, and evaluation.
Source: Department of the Navy fiscal year 2013 budget estimates.

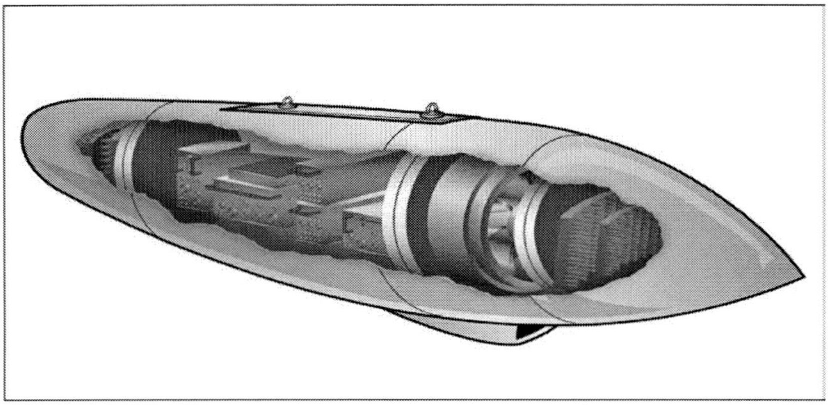

Source: NAVAIR PMA-234.

Figure 11. Next Generation Jammer.

Estimated fielding date: 2020 (Mid-band on EA-18G)

Mission description: The Next Generation Jammer will be an electronic warfare system to support the suppression of enemy air defenses, replacing and improving the capability currently provided by AN/ALQ-99 Tactical Jamming System. The Navy's EA-18G will employ the Next Generation Jammer as the electronic attack payload. In a separate increment of capability, the Navy plans to integrate the Next Generation Jammer onto the F-35B, which will eventually replace Marine Corps EA-6B Prowlers. Each increment of capability will be divided into developmental blocks—Block 1 for mid-band, Block 2 for low-band, and Block 3 for high-band frequencies.

Status: The Next Generation Jammer is nearing completion of technology maturation activities performed by four different contractors before the program's entry into the technology development phase. The Navy plans to enter the technology development phase in the third quarter of fiscal year 2013, with an engineering and manufacturing development contract planned for 2015. The Navy has adopted an evolutionary block approach to fielding the Next Generation Jammer. Initial operational capability for Block 1, on the EA-18G aircraft, is scheduled for 2020. The Navy expects to field Blocks 2 and 3 on the EA-18G in 2022 and 2024, respectively. Fielding dates for the F-35 increment's blocks are currently undetermined.

Budget: See the following table for budget information.

Table 14. DOD Planned Acquisition Investments for the Next Generation Jammer, Fiscal Years 2012-2017

Then-year dollars in millions							
	FY 2012	**FY 2013**	**FY 2014**	**FY 2015**	**FY 2016**	**FY 2017**	**Total**
RDT&E	$170.910	$187.024	$269.916	$321.817	$429.390	$528.777	**$1,907.834**
Procurement	-	-	-	-	-	-	-
Total	**$170.910**	**$187.024**	**$269.916**	**$321.817**	**$429.390**	**$528.777**	**$1,907.834**

Legend: RDT&E = research, development, testing, and evaluation.
Source: Department of the Navy fiscal year 2013 budget estimates.
Note: There is no procurement funding associated with the Next Generation Jammer in the fiscal year 2013 budget.

Source: Raytheon.

Figure 12. Miniature Air Launched Decoy (MALD)/Miniature Air Launched Decoy—Jammer (MALD-J).

Fielding dates:
2012 (MALD—actual)
2012 (MALD-J—estimated)

Mission description: MALD is an expendable decoy able to represent small, medium, or large aircraft in order to saturate or degrade enemy air defense systems. MALD-J is a variant of MALD that adds jamming capability to the decoy and forms the stand-in jamming component for the airborne electronic attack family of systems. The Air Force plans to acquire a total quantity of 596 MALD and 2,404 MALD-J units.

Status: The Air Force approved MALD for low rate initial production in 2008. The Air Force expected to procure 300 MALD units in low rate production before transitioning to full rate production. However, following flight testing failures in summer 2010—attributable, in part, to design issues with the fuel filter—and a later test failure in February 2011 caused by foreign object debris in the fuel line, the MALD system was decertified, and remaining initial operational testing and evaluation activities were suspended. After additional corrective actions by the program office to the MALD design, the system reentered operational testing in July 2011, with test shots fired in late August 2011. According to Air Force testing officials, during the last test shot in the August series (OT-8), the engine for one decoy never started after it detached from the host aircraft, causing that MALD unit to crash. This operational testing event was the final one scheduled for MALD, and DOD officials report that, in January 2012, the Air Force Operational Test and Evaluation Center delivered the MALD initial operational test and evaluation report assessing system performance.

As a result of MALD's testing shortfalls, the Air Force authorized additional low rate initial production purchases for MALD quantities—to the extent that the Air Force will now purchase the entire 596 unit inventory of MALD quantities under low rate initial production, without ever authorizing or achieving full rate production. Technical deficiencies and design changes during low rate initial production prevented demonstration of an efficient manufacturing capability, which in turn prevented MALD from meeting the department's criteria to enter full rate production. DOD policy states that in order for a system to receive full rate production approval, the system must (1) demonstrate control of the manufacturing process and acceptable reliability, (2) collect statistical process control data, and (3) demonstrate control and capability of other critical processes.[23] Because the MALD and MALD-J designs are identical—except for the addition of a jammer module to MALD-J—the absence of a proven manufacturing process for MALD introduces cost and schedule risk to production of MALD-J.

Deficiencies affecting the MALD vehicle have already contributed to MALD-J program delays. The MALD-J low rate initial production decision review—previously planned for September 2009—was delayed until September 2011. Operational testing has subsequently been delayed and is now expected to begin in May 2012. To mitigate this schedule delay, the Air Force has moved to compress MALD-J operational testing from 15 months to 7 months, which program officials report reflects an increase in test range priority and decrease in data turnaround time. According to DOD officials, however, test range execution issues such as aircraft and test equipment availability could potentially extend MALD-J operational testing beyond the currently projected completion date. In addition, the Air Force delayed, and later canceled, plans to develop a second increment of capability for MALD-J—one intended to provide more advanced jamming capabilities. Prior to these decisions, the Air Force's fiscal year 2012 budget submission outlined plans to budget $54.8 million in research, development, testing, and evaluation funding to MALD-J Increment II in fiscal year 2013. According to DOD, the Air Force is to provide a new plan for developing and procuring an Increment II variant of MALD-J and report to the Deputy Secretary of Defense by March 30, 2012.

Budget: See the following table for budget information.

Table 15. DOD Planned Acquisition Investments for MALD/MALD-J, Fiscal Years 2012-2017

Then-year dollars in millions							
	FY 2012	**FY 2013**	**FY 2014**	**FY 2015**	**FY 2016**	**FY 2017**	**Total**
RDT&E	$14.917	$0.000	$0.000	$0.000	$0.000	$0.000	**$14.917**
Procurement	83.022	87.556	89.348	92.448	94.987	95.059	**$542.420**
Total	**$97.939**	**$87.556**	**$89.348**	**$92.448**	**$94.987**	**$95.059**	**$557.337**

Legend: RDT&E = research, development, testing, and evaluation.

Source: Department of the Air Force fiscal year 2013 budget estimates.

Note: RDT&E data are for MALD-J Increment II only. There is no RDT&E funding for MALD or MALD-J.

Source: Lockheed Martin.

Figure 13. F-35 Lightning II (Joint Strike Fighter).

Estimated fielding date: To be determined

Mission description: The F-35 Joint Strike Fighter is a family of fifth-generation strike aircraft to replace and complement existing Navy, Air Force, and Marine Corps aircraft, such as the F-16 and the F/A-18. The F-35, along with the F-22A, is expected to fulfill DOD's requirement for penetrating escort jamming capability.

Status: The F-35 program entered low rate initial production in 2007, with a planned baseline acquisition of 2,886 aircraft. The program experienced development challenges, including delays in testing, leading to a program-wide review. Based on this review, DOD restructured the program in 2010, increasing the time and funding for development. This restructure triggered a breach of the critical Nunn-McCurdy cost growth threshold. Presently, the program plans to procure 2,457 aircraft, and the services are still reviewing scheduled plans for operational capability and fielding.

Budget: See the following table for budget information.

Table 16. DOD Planned Acquisition Investments for the F-35 Lightning II, Fiscal Years 2012-2017

Then-year dollars in millions							
	FY 2012	**FY 2013**	**FY 2014**	**FY 2015**	**FY 2016**	**FY 2017**	**Total**
RDT&E	$2,708.228	$2,699.498	$2,464.703	$1,899.685	$1,426.668	$1,075.495	**$12,274.277**
Procurement	6,334.916	6,149.445	6,310.537	7,786.763	9,927.117	11,207.769	**$47,716.547**
Total	**$9,043.144**	**$8,848.943**	**$8,775.240**	**$9,686.448**	**$11,353.785**	**$12,283.264**	**$59,990.824**

Legend: RDT&E = research, development, testing, and evaluation.

Source: DOD fiscal year 2013 budget estimates.

Note: The above budget figures do not include $31.874 million in fiscal year 2012 RDT&E funds and $31.748 in fiscal year 2013 RDT&E funds for the Air Force Aircraft Engine Component Improvement Program.

End Notes

[1] Irregular warfare is defined as a violent struggle among state and nonstate actors for legitimacy and influence over the relevant population(s). Irregular warfare favors indirect and asymmetric (dissimilar) approaches, though it may employ the full range of military and other capacities, in order to erode an adversary's power, influence, and will.

[2] Potential near-peer adversaries can be defined to include countries capable of waging large-scale conventional war on the United States. These nation-states can be characterized as having nearly comparable diplomatic, informational, military, and economic capacity to the United States.

[3] Anti-access threats can be defined as those that impede the deployment of U.S. forces into the combat theater, limit the locations from which those forces could effectively operate, or force them to operate from locations farther from the locus of conflict than they would normally prefer. High-value targets are persons or resources that an enemy commander requires for the successful completion of a mission.

[4] Pub. L. No. 111-84, § 1053 (a) (2009).

[5] Pub. L. No. 111-84, § 1053 (b) (2009).

[6] GAO, *Tactical Aircraft: DOD's Ability to Meet Future Requirements Is Uncertain, with Key Analyses Needed to Inform Upcoming Investment Decisions*, GAO-10-789 (Washington, D.C.: July 29, 2010).

[7] Investment total includes nearly $1.1 billion in funding for aircraft self-protection systems, which airborne electronic attack aircraft rely upon to conduct missions.

[8] GAO, *Defense Acquisitions: Assessments of Selected Weapon Programs*, GAO-11-233SP (Washington, D.C.: Mar. 29, 2011).

[9] GAO, *Best Practices: DOD Can Achieve Better Outcomes by Standardizing the Way Manufacturing Risks Are Managed*, GAO-10-439 (Washington, D.C.: Apr. 22, 2010).

[10] Pursuant to DOD Instruction 5000.02 dated December 8, 2008, low rate initial production phase is intended to ensure adequate and efficient manufacturing capability and to produce the minimum quantity necessary to provide production or production-representative articles for initial operational testing and evaluation. In the case of MALD, technical deficiencies and design changes during low rate initial production prevented demonstration of an efficient manufacturing capability, which in turn prevented the system from meeting the department's criteria to enter full rate production. Department policy further states that in order for a system to receive full rate production approval, (1) demonstrated control of the manufacturing process and acceptable reliability, (2) the collection of statistical process control data, and (3) demonstrated control and capability of other critical processes must be shown.

[11] According to DOD officials, any retrofits (design fixes) are under Raytheon (prime contractor) warranty with no additional cost to the government.

[12] According to DOD officials, the Navy negotiated a firm fixed price production contract for IDECM Block 4, under which cost risk associated with retrofits is borne by the contractor, without financial burden to the government.

[13] GAO, *Warfighter Support: DOD's Urgent Needs Processes Need a More Comprehensive Approach and Evaluation for Potential Consolidation*, GAO-11-273 (Washington, D.C.: Mar. 1, 2011).

[14] According to DOD, the Air Force is to provide a new plan for developing and procuring an Increment II variant of MALD-J and report to the Deputy Secretary of Defense by March 30, 2012.

[15] Naval Research Advisory Committee, *Science and Technology for Naval Warfare 2015-2030*, NRAC-05-3 (Arlington, Va.: August 2005).

[16] Defense Science Board, *2006 Summer Study on 21st Century Strategic Technology Vectors* (Washington, D.C.: Feb. 1, 2007).

[17] Naval Research Advisory Committee, *Status and Future of the Naval R&D Establishment* (Arlington, Va.: September 2010).

[18] The report characterized long-term naval capabilities as the "Navy-After-Next." Navy-After-Next represented concepts, platforms, and systems that had yet to be conceived, defined, or both, and for which there was no program of record.

[19] Testimony of the Honorable Zachary J. Lemnios, Assistant Secretary of Defense for Research and Engineering, in a hearing before the House Committee on Armed Services, Subcommittee on Emerging Threats and Capabilities, on March 1, 2011.

[20] Office of the Deputy Under Secretary of Defense for Acquisition and Technology, Systems and Software Engineering, *Systems Engineering Guide for Systems of Systems, Version 1.0* (Washington, D.C.: August 2008).

[21] As agreed upon with our congressional requesters, this report does not evaluate ground- or ship-based electronic warfare systems.

[22] The F/A-18C/D will not be equipped with IDECM's off-board jamming components (towed decoys) because these aircraft lack the necessary infrastructure to support these components.

[23] DOD Instruction 5000.02, *Operation of the Defense Acquisition System* (Dec. 8, 2008).

INDEX

Q

R

S

T

U

V

W